Molecular Genetics of
Early Human Development

The HUMAN MOLECULAR GENETICS series

Series Advisors

D.N. Cooper, *Institute of Medical Genetics, University of Wales College of Medicine, Cardiff, UK*

S.E. Humphries, *Division of Cardiovascular Genetics, University College London Medical School, London, UK*

T. Strachan, *Department of Human Genetics, University of Newcastle upon Tyne, Newcastle upon Tyne, UK*

Human Gene Mutation
From Genotype to Phenotype
Functional Analysis of the Human Genome
Molecular Genetics of Cancer
Environmental Mutagenesis
HLA and MHC: Genes, Molecules and Function
Human Genome Evolution
Gene Therapy
Molecular Endocrinology
Venous Thrombosis: from Genes to Clinical Medicine
Protein Dysfunction in Human Genetic Disease
Molecular Genetics of Early Human Development

Forthcoming titles
Neurofibromatosis Type 1: from Genotype to Phenotype
Analysis of Triplet Repeat Disorders

Molecular Genetics of Early Human Development

T. Strachan, S. Lindsay and D.I. Wilson
Department of Human Genetics, University of Newcastle upon Tyne, Newcastle upon Tyne, UK

βIOS
SCIENTIFIC
PUBLISHERS

© BIOS Scientific Publishers Limited, 1997

First published in 1997

A CIP catalogue record for this book is available from the British Library.

ISBN 1 859960 31 6

BIOS Scientific Publishers Ltd
9 Newtec Place, Magdalen Road, Oxford OX4 1RE, UK.
Tel. +44 (0) 1865 726286. Fax. +44 (0) 1865 246823
World-Wide Web home page: http://www.Bookshop.co.uk/BIOS/

DISTRIBUTORS

Australia and New Zealand
Blackwell Science Asia
54 University Street
Carlton, South Victoria 3053

India
Viva Books Private Limited
4325/3 Ansari Road, Daryaganj
New Delhi 110002

Singapore and South East Asia
Toppan Company (S) PTE Ltd
38 Liu Fang Road, Jurong
Singapore 2262

USA and Canada
BIOS Scientific Publishers
PO Box 605, Herndon,
VA 20172-0605

Production Editor: Priscilla Goldby
Typeset by Saxon Graphics Ltd, Derby, UK.
Printed by Information Press Ltd, Eynsham, Oxon, UK.

Contents

Contributors

Abitbol, M. Centre de Recherche Therapeutique en Ophtalmologie (CERTO), Faculté de Médecine Necker, 156 rue de Vaugirard, 75015 Paris, France

Aiton, J.F. School of Biological Sciences, Bute Building, University of St Andrews, St Andrews KY16 9TS, UK

Baldock, R.A. MRC Human Genetics Unit, Western General Hospital, Crewe Road, Edinburgh EH4 2XU, UK

Bennett, R., The Centre for Social Ethics and Policy, Humanities Building, University of Manchester, Oxford Road, Manchester M13 9PL, UK

Bolton, V. Assisted Conception Unit, Department of Obstetrics and Gynaecology, King's College School of Medicine and Dentistry, Denmark Hill, London SE5 8RX, UK

Boncinelli, E. Dipartimento di Ricerca e Tecnologica, Istituto Scientifico H San Raffaele, Via Olgettina 58, 20132 Milan, Italy

Bullen, P. Molecular Genetics Unit, Department of Human Genetics, University of Newcastle upon Tyne, Ridley Building, Claremont Place, Newcastle upon Tyne NE1 7RU, UK

Daniels, R. Molecular Embryology Unit, Institute of Child Health, 30 Guilford Street, London WC1N 1EH, UK

Davidson, D.R. MRC Human Genetics Unit, Western General Hospital, Crewe Road, Edinburgh EH4 2XU, UK

Fitzpatrick, D.R. Molecular Medicine Centre, Western General Hospital, Crewe Road, Edinburgh EH4 2XU, UK

Harris, J. The Centre for Social Ethics and Policy, Humanities Building, University of Manchester, Oxford Road, Manchester M13 9PL, UK

Kaufman, M. Department of Anatomy, University of Edinburgh Medical School, Teviot Place, Edinburgh EH8 9AG, UK

Lako, M. Molecular Genetics Unit, Department of Human Genetics, University of Newcastle upon Tyne, Ridley Building, Claremont Place, Newcastle upon Tyne NE1 7RU, UK

Lindsay, S. Molecular Genetics Unit, Department of Human Genetics, University of Newcastle upon Tyne, Ridley Building, Claremont Place, Newcastle upon Tyne NE1 7RU, UK

McLachlan, J.C. School of Biological Sciences, Bute Building, University of St Andrews, St Andrews KY16 9TS, UK

Monk, M. Molecular Embryology Unit, Institute of Child Health, 30 Guilford Street, London WC1N 1EH, UK

Rankin, J. Molecular Genetics Unit, Department of Human Genetics, University of Newcastle upon Tyne, Ridley Building, Claremont Place, Newcastle upon Tyne NE1 7RU, UK

Robson, S.C. Fetal Medicine Unit, Leazes Wing, Royal Victoria Infirmary, Newcastle upon Tyne NE1 4LP, UK

Smart, S.D. School of Biological Sciences, Bute Building, University of St Andrews, St Andrews KY16 9TS, UK

Strachan, T. Molecular Genetics Unit, Department of Human Genetics, University of Newcastle upon Tyne, Ridley Building, Claremont Place, Newcastle upon Tyne NE1 7RU, UK

Thorogood, P. Developmental Biology Unit, Institute of Child Health, 30 Guilford Street, London WC1N 1EH, UK

Whiten, S.C. School of Biological Sciences, Bute Building, University of St Andrews, St Andrews KY16 9TS, UK

Wilson, D.I. Molecular Genetics Unit, Department of Human Genetics, University of Newcastle upon Tyne, Ridley Building, Claremont Place, Newcastle upon Tyne NE1 7RU, UK

Abbreviations

α-MEM	alpha minimal essential medium
3-D	three-dimensional
AER	apical ectodermal ridge
AMH	anti-Müllerian hormone
CNS	central nervous system
CRL	crown-rump length
CSF	colony-stimulating factor
CSFM	complex serum-free medium
CVS	chorionic villus sampling
DEPC	diethylpyrocarbonate
dNTP	deoxynucleotide triphosphates
DTT	dithiothreitol
DZ	dizygous (twinning)
E14	embryonic day 14
EBSS	Earle's balanced salt solution
ECM	extracellular matrix
EDTA	ethylenediaminetetraacetic acid
EGF	epidermal growth factor
EGF-R	epidermal growth factor receptor
ELISA	enzyme-linked immunosorbent assays
EM	electron microscopy
ES	embryonic stem (cells)
EST	expressed sequence tag
FBS	fetal bovine serum
FMRP	FMR1 protein
FSH	follicle stimulating hormone
gpBO	glycoprotein BO
hCG	human chorionic gonadotrophin
HCS	human cord serum
HOM-C	homeotic complex
hprt	hypoxanthine phosphoribosyl transferase
HTF	human tubal fluid (medium)
ICM	inner cell mass
IGF	insulin-like growth factor
IL	interleukin
IVF	*in-vitro* fertilization
KH	ribonucleoprotein K homology
LH	luteinizing hormone
LIF	leukaemia-inhibitory factor
LIF-R	leukaemia-inhibitory factor receptor
MGD	mouse genome database
MGEIR	mouse gene expression information resource

MPK	myotonin protein kinase
MZ	monozygous (twinning)
NMR	nuclear magnetic resonance
NTD	neural tube defects
OMFL	oromandibulofacial limb hypogenesis
p.c.	post conception
PAF	platelet-activating factor
PAGE	polyacrylamide gel electrophoresis
PBS	phosphate-buffered saline
PCC	premature chromosome condensation
PDGF	platelet-derived growth factor
PFA	paraformaldehyde
QTVR	quick time virtual reality
RAR	retinoic acid receptor
RARE	retinoic acid response element
RPE	retinal pigmentary epithelium
RT	reverse transcriptase
RT-PCR	reverse transcriptase-polymerase chain reaction
RXR	retinoid X receptor
SAGE	serial analysis of gene expression
SCF	stem cell factor
SDS	sodium dodecyl sulphate
SEM	scanning electron microscopy
SNuPE	single nucleotide primer extension
SP-1	(pregnancy-) specific β_1-glycoprotein
SSC	standard saline citrate
T6	Tyrode's 6 (medium)
TE	trophectoderm
TGF-α	transforming growth factor-α (also β)
TNF-α	tumour necrosis factor-α
TOP	termination of pregnancy
VR	virtual reality
YAC	yeast artificial chromosome

Preface

We believe that the present volume is the first of its kind. The study of early human development has traditionally been mostly limited to anatomical and morphological studies of retrieved embryonic or fetal material following termination of pregnancy. Descriptive embryological studies of this type have long been conducted on archived human embryo collections, providing valuable information on our early development. But for fundamental understanding of the mechanisms of vertebrate development we have relied almost totally on studies of animal models of development such as the mouse, chick, *Xenopus* and zebrafish. Valuable although such models undoubtedly are, we are becoming increasingly aware of species differences that limit the extent to which we can extrapolate from animal models to early human development. The number of genes identified as playing a role in human developmental disorders is rapidly increasing and so from the clinical viewpoint, too, there is an increasing need for information on normal human development to provide a framework for understanding abnormal human development. Molecular genetic studies on human embryonic material are providing a new dimension to understanding early human development and the present volume is an attempt to bring together various facets of this field.

The first molecular genetic studies to chart gene expression in early human development were limited to low-resolution methods. In the 1980s, methods such as RT-PCR, for example, began to be applied to study gene expression in preimplantation human embryos accessed through *in vitro* fertilization programmes. In the early 1990s, the first high-resolution gene expression studies in postimplantation human embryos were published, opening up a new field of human genome analysis. Now, there is an increasing realization in the human genetics community that, wherever possible, we should be studying gene expression directly in human cells, instead of simply relying on extrapolation from animal models. Because of ethical and practical considerations, however, certain stages of early human development will not be easily accessible and animal models will always be important reference points. Referencing gene expression patterns in early human development against those in animal models will help us to appreciate the degree of species differences in gene expression. In addition, studies of gene expression in human embryonic tissue should provide molecular genetic landmarks of development which could be expected to illuminate our understanding of early development. This will be particularly important in areas where there are comparatively few anatomical landmarks, such as in the developing human brain. We look forward to an exciting future for this developing new field.

Tom Strachan
Susan Lindsay
David I. Wilson
(*Newcastle upon Tyne*)

Acknowledgements

The colour images of human embryos in Chapter 3 have been included with the generous support of Carl Zeiss and Imaging Associates, and we are grateful for their help.

Why study human embryos? The clinical need

David R. FitzPatrick

1. Introduction

The transition from embryonic to fetal life in humans is generally considered to occur at 8 postovulatory weeks. This is an arbitrary but useful boundary which is based on the fact that approximately 90% of recognizable adult body structures have formed by this stage (O'Rahilly, 1979). Whatever the exact definition, human embryogenesis is undoubtedly one of the most exciting and hazardous periods of human existence. It begins with a single pluripotent cell approximately 100 μm in diameter and ends approximately 56 days later as a 2.9 cm multicellular organism with its future body shape almost complete, deriving all of its nourishment from a highly efficient, parasitic apparatus, the placenta. This remarkable period of growth, shown graphically in *Figure 1*, is accompanied by a panoply of cellular activity including: differentiation, syncytiation, migration and programmed death.

Most pregnancies that last into the third trimester have survived the embryonic period more or less intact. However, many are not so lucky, with 20–30% of clinically detectable human pregnancies lost in the first trimester (Ellish *et al.*, 1996; Zinaman *et al.*, 1996). This chapter will review some of the clinical problems associated with embryogenesis and review the potential benefits of studying human embryogenesis.

2. What can go wrong in human embryogenesis?

2.1 Describing human embryos

In obstetric practice the duration of pregnancy is most often staged as an estimate of time from ovulation judged by the last menstrual period. This method is not sufficient for embryological use and George Streeter (1873–1948), working at the Carnegie Laboratories, devised an anatomical staging system for human embryos. This system has recently been revised and updated (O'Rahilly and Muller, 1987)

Molecular Genetics of Early Human Development, edited by T. Strachan, S. Lindsay and D.I. Wilson.
© 1997 BIOS Scientific Publishers Ltd, Oxford.

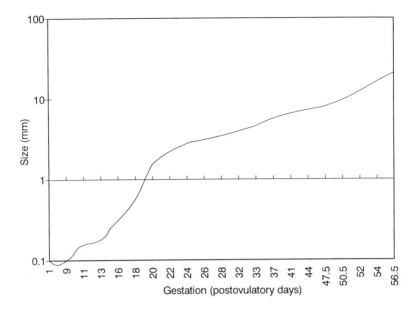

Figure 1. Growth of the human embryo. A graphical representation of human embryonic growth using the logarithmic scale. The measurements (mm) are derived from the postfixation maximum length estimations of embryos in the Carnegie collection (O'Rahilly and Muller, 1987). See Dickey and Gasser (1993) for a discussion of the limitations of applying such data *in vivo*.

and is summarized in *Table 1*. Human embryology will be comprehensively described in Chapter 6 and only the major events in embryogenesis will be briefly described here (using the Carnegie system) as they relate to embryopathology.

2.2 Fertilization, aneuploidy and aneusomy

In most human conceptions fertilization occurs by cytoplasmic penetration of the haploid male pronuclei (containing one of each autosome with either an X or Y chromosome) into the haploid oocyte (containing one of each autosome and an X chromosome) in the ampullary region of the Fallopian tubes. One of the most common embryopathies which can be dated to the time of fertilization is aneuploidy. Humans appear to be uniquely prone to parental meiotic errors, which result in the frequent occurrence of chromosomally abnormal gametes. In human sperm whole-chromosome aneuploidy occurs in 3 per 1000 sperm nuclei per chromosome (Pellestor *et al.*, 1996) with an overall rate of chromosomal abnormality of up to 20% (Bischoff *et al.*, 1994). Female gametes are obviously less available for study than sperm and the rate of aneuploidy in oocytes has evaded an unbiased assessment. Aneuploidy in oocytes does, however, appear to be correlated with maternal age and may also be as high as 20% (Roberts and O'Neill, 1995).

In vitro fertilization programmes have also enabled the cytogenetic analysis of preimplantation human embryos. A recent study of 178 embryos found that 22.5% were chromosomally abnormal (Jamieson *et al.*, 1994). Most autosomal aneuploidy

Table 1. Carnegie staging of human embryos

Carnegie Stage	Age (days)	Size (mm)	Major features
1	1	0.1	Fertilization
2	1.5	0.1	2–16 cells
3	4	0.1	Blastocyst
4	5	0.1	Attaching blastocyst
5	7	0.1	Implantation
6	13	0.2	Chorionic villi formation; primitive streak
7	16	0.4	Notochord formation
8	18	1.25	Gastrulation; neural fold formation
9	20	2	Somite formation
10	22	2.75	Start of neural tube fusion
11	24	3.5	Optic vesicle formation; anterior neuropore closes
12	26	4	Upper limb buds appear
13	28	5	Lower limb buds, lens disc and otic vesicle appear
14	32	6	Optic cup appears
15	33	8	Formation of nasal pit, hand plate and future cerebral hemispheres
16	37	9.5	Auricular hillocks and foot plate appear
17	41	12.5	Finger rays distinct; nasofrontal groove appears
18	44	15	Ossification begins; eyelid folds
19	47.5	17	Trunk elongating and straightening
20	50.5	20	Upper limbs longer and bent at elbow
21	52	23	Fingers longer; hands approach each other
22	54	25.5	Eyelids and external ears more distinct
23	56.5	29	External genitalia well developed

has an extremely deleterious effect on the embryo, which may be recognized phenotypically even in Carnegie Stages 2–4 (Kligman *et al.*, 1996). Clinically this can present as spontaneous, first trimester abortion (common outcome) or a birth with multiple physical and mental handicaps (rare outcome). Approximately 50% of spontaneously aborted pregnancies and approximately 6% of stillborn infants are aneuploid [cf approx 0.9% in livebirths (Jacobs and Hassold, 1995)]. As can be seen in *Figure 2* the probability of surviving to birth as an aneuploid fetus is highly dependent on which chromosome is involved.

While it is clear that aneuploidy is a very common human disorder, it is not clear how chromosomal anomalies induce embryopathy. Recently, Kuliev *et al.* (1996) used reverse transcriptase–polymerase chain reaction (RT-PCR) to demonstrate abnormalities in the expression of *HoxA7*, *HoxB5* and *HoxC6* genes in preimplantation embryos with several different aneuploidies. These preliminary data would support the idea that aneuploidy may produce a relatively non-specific 'bug' into the developmental programme by interfering with genes involved in pattern formation. However, further studies of gene expression in normal embryos and embryos with known aneuploidy may allow the molecular definition of chromosomal developmental pathologies.

An interesting recent addition to this group of very early embryopathies has come from the recognition of uniparental disomy for single chromosomes (Tomkins *et al.*,

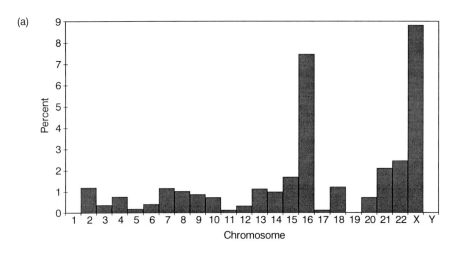

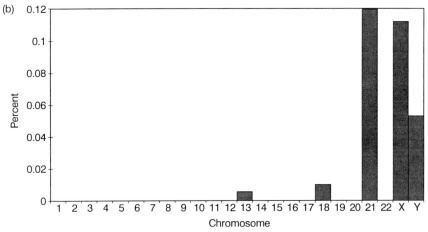

Figure 2. Whole-chromosome aneuploidy in human embryos. Graphical representations of the percentages of pregnancies affected with chromosome abnormalities (Y axis) involving each human chromosome (X axis). (a) Spectrum of whole-chromosome aneuploidy in spontaneously aborted fetuses. Note that aneuploidy is very common (34.7% overall) and every chromosome is represented apart from 1 and 19. (b) Aneuploidy is rarely seen in liveborn infants (0.3% overall) and involves only chromosomes 13, 18, 21, X and Y.

1996 and references therein) or, rarely, a whole genome (Strain *et al.*, 1995). These rare cases demonstrate the importance of the different genomic imprinting that takes place during male and female gametogenesis. For this reason the term aneusomy has been proposed to describe all chromosome abnormalities, in preference to the more specific aneuploidy. The high rates of aneusomy should be borne in mind by all who plan to study gene-expression patterns in human embryos. The rate of chromosomal abnormalities in phenotypically normal embryos is not known and multiple embryos will need to be analysed before conclusions can be drawn about the 'normal' pattern of expression of any particular gene.

2.3 Ectopic blastocyst implantation

After fertilization, mitotic cell divisions create first a ball of 16 cells (Stage 2) and then a blastocyst cavity (Stage 3) with a small mound of cells on the inner aspect of this cavity (the inner cell mass from which the embryo will develop. Until this stage there has been almost no overall growth accompanying these cell divisions. On days 5–6 (Stage 4) there is attachment of the embryo to the endometrium, and by day 7 (Stage 5) the blastocyst has implanted in the uterine wall, gaining a new source of nourishment enabling a rapid and sustained period of growth. In approximately 1.6% of all pregnancies the Stage 4 embryo implants ectopically. By far the most common site of ectopic implantation is in the Fallopian tubes, although both ovarian and abdominal pregnancies do occur. Ectopic pregnancy is usually fatal for the embryo and can be a life-threatening surgical emergency for the mother. Approximately two-thirds of cases of tubal pregnancy can be explained by known risk factors with a presumed mechanical basis (pelvic inflammatory disease, intrauterine contraceptive device, previous uterine surgery). In the remaining one-third the aetiology is unknown. In this group there is an association with maternal age and recurrent pregnancy loss, which has led to the speculation that there may be a causative role for gametic or embryo pathology (Bouyer *et al.*, 1996).

2.4 Twinning

Twinning is another human 'disorder' which dates specifically to the preimplantation stage of development. Dizygous (DZ) or fraternal twinning results from independent fertilization of two different eggs. Interestingly, the birth incidence of DZ twinning is population-dependent and ranges between 1 in 330 in oriental populations to 1 in 25 in Nigerians (1 in 120 in Caucasians). There is considerable evidence that these differences may reflect both population-specific genetic and environmental influences (Chen *et al.*, 1992; James, 1995; Picard *et al.*, 1989).

Monozygous (MZ; identical) twinning occurs in 1 in 250 births in all human populations studied and is due to the complete cleavage of the inner cell mass of the blastocyst in the preimplantation period. Although the cause or precise timing of this cleavage is not known, the application of non-linear dynamic mathematical modelling to the periodicity of MZ twinning has led to the proposal that there may be 'a limited number of intrinsic and deterministic interacting system components' (Phillipe, 1994). Previous reports have suggested that these components may have an identifiable genetic basis (Gleeson *et al.*, 1994) and evidence from *in vitro* fertilization programmes has suggested that at least one of these determinants may be a natural thinness in the zona pellucida of embryos destined to become MZ twins (Alikani *et al.*, 1994). There have also been suggestions on clinical grounds that MZ twinning in females may be related to segmentally discordant X-inactivation in the embryo, although this has not been substantiated using postnatal molecular analysis (Goodship *et al.*, 1996).

There are many reasons why twinning should be studied in humans. First, the laboratory mouse, the most popular mammalian developmental model, does not appear to produce MZ twins naturally (McLaren *et al.*, 1996). Second, the process of MZ twinning is particularly prone to errors, resulting in a high incidence of congenital malformations (Chen et al., 1992) or, rarely, conjoined twins. Third, and perhaps most importantly, the study of twins has been used widely to assess whether a disease has a genetic component to its aetiology (see Table 2). Since most MZ and DZ twins share a common environment, at least in intrauterine and early postnatal life, by comparing MZ and DZ twins where one twin has developed condition x, if more MZ co-twins also develop x (twin concordance) than DZ co-twins it suggests that there is a significant and measurable genetic contribution to the aetiology of that disorder (heritability). A better understanding of the biology of twinning may allow us to recognize the benefits and possible limitations of such studies.

Table 2. Twinning

Disorder	MZ % concordance	DZ % concordance
Peptic ulcer	53	36
Myocardial infarction (males)	39	26
Myocardial infarction (females)	44	14
Non-insulin dependent diabetes mellitus	70	15

Extracted from King et al. (1992). MZ, monozygous; DZ, dizygous.

2.5 Early pregnancy loss

After implantation the cytotrophoblast invades the endometrium and will ultimately form the placenta. By day 10 (Stage 5–6) the embryo has developed into a plate-like structure made up of two layers of cells: the ectoderm and the primitive endoderm. It is from this stage until the end of the first trimester that the period of maximum embryo wastage begins. Of the 20–30% of clinically recognizable human pregnancies that are lost during embryogenesis, approximately 50% can be explained by aneuploidy (the rate of aneusomy is not known). Although the causes of the other half of early pregnancy losses are unclear, there are undoubtedly maternal immunological and other factors (Sargent and Dokras, 1996) interacting with a possible environmental contribution (Weinberg et al., 1994). Recurrent unexplained pregnancy loss, in particular, is the cause of much suffering to affected couples and is an area of research that human embryologists as well as gynaecologists should prioritize.

2.6 Congenital malformations

Organogenesis is a general term for the process of cellular differentiation to form tissues present in the adult, and is what most people think of as the 'nuts and bolts' of embryogenesis (see Chapter 6). It involves an extraordinarily complex symphony of cellular activity and, as such, it is not surprising that errors in one or more of these processes are common in the human population. These errors present clinically as congenital malformations, which are classed as major (those with an adverse effect on either the function or social acceptability of the individual) or minor (those with neither medical nor cosmetic consequence to the patient) (Marden *et al.*, 1964). Of all births 15–30 in 1000 have major congenital abnormalities, with 10–20% of these births having multiple malformations [at least 1 in 1000 with two and 0.5 in 1000 with three malformations (Eriksson and Zetterstrom, 1994)]. The most common human major malformations in two European populations in the period 1965–1985 are shown in *Table 3* (Eriksson and Zetterstrom, 1994; Knox and Lancashire, 1991).

The aetiology of most birth defects is not known. Kalter and Warkany (1983a) estimated the proportion of major malformations caused by specific aetiologies as: monogenic 7.5%, chromosomal 6%, maternal infections 2%, maternal diabetes 1.5%, anticonvulsant drugs 1.5%. Thus < 20% of these defects have identifiable causes. It is likely that gene–environment interactions will explain a significant proportion of the remaining defects. An example of how such interactions may be dissected is found in the work leading to the discovery of an association between neural tube defects, mild maternal hyperhomocysteinaemia and a common point mutation in the methylenetetrahydrofolate reductase gene (van der Put *et al.*, 1995 and refs therein). This finding, at least partially, explains why periconceptual supplementation with folic acid prevents recurrence of neural tube defects [NTD (MRC Vitamin Study Research Group, 1991)].

Table 3. The incidence of congenital malformations in European populations

Malformation	Incidence per 1000 births
Talipes	5.6
Neural tube defects[a]	3.2
Cardiac defects	1.5–3.9
Congenital dislocation of the hip	2.0
Polydactyly/syndactyly	2.0
Hypospadias	1.5
Facial clefts	1.2–1.6
Down's syndrome	1.2–1.4
Hydrocephalus	0.9
Anal atresia	0.6
Eye defects	0.4
Limb reduction	0.4
Tracheo-oesophageal fistula	0.3
Renal agenesis	0.2–0.3
Craniosynostosis	0.15
Gastroschisis	0.1

Taken from Knox and Lancashire (1991) and Eriksson and Zetterstrom (1994).
[a]Rate prior to widescale prenatal diagnosis or folate supplementation.

3. Understanding errors of morphogenesis

3.1 Epidemiological approaches

Our understanding of the aetiology and prevention of birth defects such as NTD owes much to epidemiological research. Indeed, many of the first clues to successful primary prevention of congenital malformations come from population surveillance studies, such as on the teratogenic effect of prescription drugs (Correy *et al.*, 1991; Kalter and Warkany, 1983b) or the probable detrimental effect of particular environmental factors (Rothman *et al.*, 1995). Such studies open the way to public health campaigns which may make a significant impact on the incidence of birth defects. However, basic researchers must work with epidemiologists to determine the molecular basis of such phenomena as such information may then allow further, more targeted preventative strategies to be developed. In particular, the aetiology of malformations in the offspring of diabetic mothers (Janssen *et al.*, 1996) and mothers with epilepsy (Andermann *et al.*, 1995) deserve further study as the risk factors have been identified but successful primary prevention will require a better understanding of the molecular basis of the embryopathy.

3.2 Genetic approaches

Most human malformations show at least some evidence of familial clustering (Ananijevicpandey *et al.*, 1992) and although few 'Mendelise' (Wilkie *et al.*, 1994), there is supporting evidence for a genetic contribution to their aetiology (Jain *et al.*, 1993). This familial clustering enables the use of either candidate gene or genome-wide haplotype association searches to identify potentially causative loci. The former has been used successfully in the study of orofacial clefting (Hwang *et al.*, 1995). The latter has been used successfully in other complex genetic traits (Todd and Farrall, 1996) but has not, as yet, been widely applied to the study of particular malformations. The identification of candidate genes at each of these loci by genome-wide searches will be considerably aided by the continuing work of the human genome project (http: //weber.u.washington.edu/~roach/human_ genome_progress2.html). Since genetic epidemiology predicts that there are likely to be multiple causative loci for each defect studied (FitzPatrick and Farrell, 1993), it is unlikely that there will be any shortage of interesting genes for human embryologists to study. This combination of 'big science', molecular epidemiology with molecular embryology is likely to be very productive in the future.

3.3 Animal models

Much of the progress in our understanding of morphogenesis has been through innovative manipulation of two well-studied invertebrate model animals: the roundworm (*Caenorrhabditis elegans* http: //dauerdigs.biosci.missouri.edu/Dauer-World/Wormintro.html) and the fruit fly (*Drosophila melanogaster* http: //www-leland.stanford.edu/%7Eger/drosophila.html). There are also four vertebrate animal model systems that have been very productive: African clawed toad (*Xenopus laevis* http: //vize222.zo.utexas.edu/), zebrafish (*Danio rerio* http: //zfish.uoregon. edu/), chick (*Gallus domesticus* http://ppc30.bio.ucalgary.ca/cmmb403/tutorial TOC.html) and mouse (*Mus musculus* http: //glengoyne.hgu.mrc.ac.uk/Genex/ Documentation/). The particular strengths of each of these models are summarized in *Table 4* (the mouse as a model will be discussed in detail in Chapter 2). The major advantages that these animals have over human embryos are the ability to generate and propagate

Table 4. Animal models in developmental biology

Organism	Adult size	Generation time	Strengths
Caenorrhabditis elegans	~ 1 mm	3 days	A complete fate map is available; many mutants and a near-complete genome project
Drosophila melanogaster	3 mm	10 days	Classic genetic model; enormous number of mutants available
Xenopus laevis	80 mm	26 weeks	Large zygote, externally fertilized and easily manipulable; excellent for the study of mesoderm induction and patterning
Danio rerio	30 mm	12 weeks	Transparent embryos; large-scale mutagenesis project complete
Gallus domesticus	200 mm	26 weeks	Externally fertilized; visible embryology; tissue transplantation and xenografting possible; limb, heart and facial development particularly well studied
Mus musculus	60 mm	10 weeks	See Chapter 2

mutants on a stable genetic background, and the ability to manufacture transgenic animals and perform tissue transplantation and it is likely that most of the major questions in morphogenesis will be answered using such model systems. This work will generate many potential candidate genes for human malformation and identification of human orthologues and their chromosomal location will, in most cases, be a computer-based activity. The major disadvantage of animal models is, simply, that they are not human and all clinical relevance must be implied. Therefore, before money and time is wasted on a gene with an interesting gene-expression pattern in mouse embryos, the gene activity in human embryos must be studied.

4. Therapeutic considerations

There has been considerable scientific (Ansari and Sundstrom, 1996) and ethical (Vawter and Gervais, 1995) interest generated by the use of human embryonic and fetal cells in tissue transplantation. Such transplants have now been reported in the treatment of Parkinson's disease (Jacques *et al.*, 1995), Huntington's disease (Kurth *et al.*, 1996), insulin-dependent diabetes mellitus (Dordevic *et al.*, 1995), retinitis pigmentosa (Delcerro *et al.*, 1995) and various haematological disorders (Sukhikh *et al.*, 1994). These therapies assume the existence of stem cells in the transplanted tissue which will then terminally differentiate into the deficient tissue in the host organ. An improved understanding of developmental processes, together with the ability to isolate and propagate stem cells, may have far wider implications in transplantation biology. It is perhaps not too far-fetched to speculate that in the future we may be able to induce whole-organ growth from cultured human stem cells, for example of a kidney, which may revolutionize the treatment of chronic organ failure.

5. Conclusion

The last 20 years has seen the emergence of developmental biology as one of the most exciting areas of modern biomedical research. It should be remembered that the ethical issues raised by the use of human embryos for research purposes will require careful and continuous consideration (see Chapter 4; Annas *et al.*, 1996; Marshall, 1994). It is clear that all genetic alteration or most experimental manipulation of live human embryos is (and will remain for the foreseeable future) completely unacceptable to both the scientific community and the wider public. For these reasons the human will never challenge the mouse as the mammalian model system in experimental embryology.

It may, at first glance, seem that the human embryo should be relegated to the rather tranquil scientific backwater of comparative mammalian embryology. Humans do have the advantage of a large number of naturally occurring, well-studied developmental pathologies. There are approximately 140 different (and rare) Mendelian loci causing congenital malformations (Wilkie *et al.*, 1994), as well as a large number of living individuals with single-site malformations, associations, developmental anomalies and syndromes with more complex aetiologies. This resource alone will generate many genes to be studied in embryos over the next few years.

Why is it important to study human embryos? The answer is because humans are not worms, flies, birds, toads, fish or rodents but humans. It must be hoped that by integrating the important information generated by these animal model systems, human genetic and epidemiological studies combined with descriptive studies of gene expression during specific periods of morphogenesis in human embryos we may understand, and ultimately prevent, socially, medically and economically important human diseases.

References

Alikani M, Noyes N, Cohen J, Rosenwaks Z. (1994) Monozygotic twinning in the human is associated with the zona pellucida architecture. *Hum. Reprod.* 9: 1318–1321.
Ananijevicpandey J, Jarebinski M, Kastratovic B, Vlajinac H, Radojkovic Z, Brankovic D. (1992) Case control study of congenital malformations. *Eur. J. Epidemiol.* 8: 871–874.
Andermann E, Kaneko S, Battino D, Goto M. (1995) Prospective international multicenter study on congenital malformations in offspring of mothers with epilepsy. *Am. J. Hum. Genet.* 57: 905.
Annas GJ, Caplan A, Elias S. (1996) The politics of human embryo research — avoiding ethical gridlock. *N. Engl. J. Med.* 334: 1329–1332.
Ansari AA, Sundstrom JB. (1996) Transplantation of fetal tissues. *Immunol. Allergy Clin. N. Am.* 16: 333.
Bischoff FZ, Nguyen DD, Burt KJ, Shaffer LG. (1994) Estimates of aneuploidy using multicolor fluorescence in-situ hybridization on human sperm. *Cytogenet. Cell Genet.* 66: 237–243.
Bouyer J, Tharauxdeneux C, Coste J, Jobspira N. (1996) Ectopic pregnancy — risk factors related to egg anomalies. *Rev. Epidemiol. Sante Publique* 44: 101–110.
Chen CJ, Wang CJ, Yu MW, Lee TK. (1992) Perinatal mortality and prevalence of major congenital malformations of twins in Taipei city. *Acta Genet. Med. Gemellol.* 41: 197–203.
Chen CJ, Lee TK, Wang CJ, and Yu MW. (1992) Secular trend and associated factors of twinning in Taiwan. *Acta Genet. Med. Gemellol.* 41: 205–213.
Correy JF, Newman NM, Collins JA, Burrows EA, Burrows RF, Curran JT. (1991) Use of prescription drugs in the 1st trimester and congenital malformations. *Aust. NZ J. Obstet. Gynaecol.* 31: 340–344.

Delcerro M, Das T, Reddy VL, Diloreto D, Jalali S, Little C, Delcerro C, Rao GN, Sreedharan A. (1995) Human fetal neural retinal cell transplantation in retinitis pigmentosa. *Vision Res.* **35**: 3336.

Dickey RP, Gasser RF. (1993) Computer analysis of the human embryo growth curve — differences between published ultrasound findings on living embryos in-utero and data on fixed specimens. *Anat. Rec.* **237**: 400–407.

Dordevic PB, Lalic NM, Brkic S et al. (1995) Human fetal islet transplantation in iddm patients – an 8-year experience. *Transplant. Proc.* **27**: 3146–3147.

Ellish NJ, Saboda K, O'Connor J, Nasca PC, Stanek EJ, Boyle C. (1996) A prospective study of early pregnancy loss. *Hum. Reprod.* **11**: 406–412.

Eriksson M, Zetterstrom R. (1994) Environment and epidemiology of congenital malformations. *Acta Paediatr.* **83**: 30–34.

FitzPatrick DR, Farrall M. (1993) An estimation of the number of susceptibility loci for isolated cleft palate. *J. Craniofac. Genet. Dev. Biol.* **13**: 230–235.

Gleeson SK, Clark AB, Dugatkin LA. (1994) Monozygotic twinning — an evolutionary hypothesis. *Proc. Natl. Acad. Sci. USA* **91**: 11363–11367.

Goodship J, Carter J, Burn J. (1996) X-inactivation patterns in monozygotic and dizygotic female twins. *Am. J. Med. Genet.* **61**: 205–208.

Hwang SJ, Beaty TH, Panny SR, Street NA, Joseph JM, Gordon S, McIntosh I, Francomano CA. (1995) Association study of transforming growth-factor-alpha (tgf-alpha) taqi polymorphism and oral clefts — indication of gene–environment interaction in a population-based sample of infants with birth defects. *Am. J. Epidemiol.* **141**: 629–636.

Jacobs PA, Hassold TJ. (1995) The origins of numerical chromosomal abnormalities. *Adv. Genet.* **33**: 101–133.

Jacques DB, Kopyov OV, Markham CH, Lieberman A, Snow B, Vingerhoets F. (1995). 3-year experience of fetal mesencephalic tissue transplantation in Parkinson patients. *Exp. Neurol.* **135**: 166.

Jain VK, Nalini P, Chandra R, Srinivasan S. (1993) Congenital malformations, reproductive wastage and consanguineous mating. *Aust. NZ J. Obstet. Gynaecol.* **33**: 33–36.

James WH. (1995) Are natural twinning rates continuing to decline? *Hum. Reprod.* **10**: 3042–3044.

Jamieson ME, Coutts JRT, Connor JM. (1994) The chromosome constitution of human preimplantation embryos fertilized in-vitro. *Hum. Reprod.* **9**: 709–715.

Janssen PA, Rothman I, Schwartz SM. (1996) Congenital malformations in newborns of women with established and gestational diabetes in Washington-state, 1984–91. *Paediatr. Perinat. Epidemiol.* **10**: 52–63.

Kalter H, Warkany J. (1983a) Congenital malformations: etiological factors and their role in prevention I. *N. Engl. J. Med.* **308**: 424–431.

Kalter H, Warkany J. (1983b) Congenital malformations: etiological factors and their role in prevention II. *N. Engl. J. Med.* **308**: 491–497.

King RA, Rotter JI, Motulsky AG. (1992) *The Genetic Basis of Common Diseases.* Oxford University Press, Oxford.

Kligman I, Benadiva C, Alikani M, Munne S. (1996) The presence of multinucleated blastomeres in human embryos is correlated with chromosomal abnormalities. *Hum. Reprod.* **11**: 1492–1498.

Knox EG, Lancashire RJ. (1991) *Epidemiology of Congenital Malformations.* London.

Kuliev A, Kukharenko V, Morozov G et al. (1996) Expression of homeobox-containing genes in human preimplantation development and in embryos with chromosomal aneuploidies. *J. Assist. Reprod. Genet.* **13**: 177–181.

Kurth MC, Kopyov O, Jacques DB. (1996) Improvement in motor function after fetal transplantation in a patient with Huntington's disease. *Neurology* **46**: 22006.

Marden PM, Smith DW, McDonald MJ. (1964) Congenital anomalies in the newborn infant including minor variations. *J. Pediatr.* **64**: 357–371.

Marshall E. (1994) Human embryo research — Clinton rules out some studies. *Science* **266**: 1634–1635.

McLaren A, Molland P, Signer E. (1996) Does monozygotic twinning occur in mice? *Genet. Res.* **66**: 195–202.

MRC Vitamin Study Research Group (1991) Prevention of neural tube defects, results of the Medical Research Council vitamin study. *Lancet* **338**: 131–137.

O'Rahilly R. (1979) Early human development and the chief sources of information on staged human embryos. *Eur. J. Obstet. Gynec. Reprod. Biol.* **9**: 273–280.

O'Rahilly R, Muller F. (1987) *Developmental Stages in Human Embryos.* Carnegie Institute, Washington.

Pellestor F, Girardet A, Coignet L, Andreo B, Charlieu JP. (1996) Assessment of aneuploidy for chromosome-8, chromosome-9, chromosome-13, chromosome-16, and chromosome-21 in human sperm by using primed in-situ labeling technique. *Am. J. Hum. Genet.* **58**: 797–802.

Philippe P. (1994) MZ twinning — chance or determinism — an essay in nonlinear dynamics (chaos). *Ann. Hum. Biol.* **21**: 423–434.

Picard R, Fraser D, Hagay ZJ, Leiberman JR. (1989) Twinning in southern Israel — secular trends, ethnic variation and effects of maternal age and parity. *Eur. J. Obstet. Gynecol. Reprod. Biol.* **33**: 131–139.

Roberts CG, O'Neill CO. (1995) Increase in the rate of diploidy with maternal age in unfertilized in-vitro fertilization oocytes. *Hum. Reprod.* **10**: 2139–2141.

Rothman KJ, Moore LL, Singer MR, Nguyen UDT, Mannino S, Milunsky A. (1995) Teratogenicity of high vitamin A intake. *N. Engl. J. Med.* **333**: 1369–1373.

Sargent IL, Dokras A. (1996) Embryotoxicity as a marker for recurrent pregnancy loss. *Am. J. Reprod. Immunol.* **35**: 383–387.

Strain L, Warner JP, Johnston T, Bonthron DT. (1995) A human parthenogenic chimera. *Nature Genetics* **11**: 164–169.

Sukhikh GT, Molnar EM, Malaitsev VV, Bogdanova IM. (1994) Transplantation of human fetal tissue in hematology. *Bull. Exp. Biol. Med.* **117**: 376–378.

Todd JA, Farrall M. (1996) Panning for gold — genome-wide scanning for linkage in type-1 diabetes. *Hum. Mol. Genet.* **5**: 1443–1448.

Tomkins DJ, Roux A-F, Waye J, Freeman VCP, Cox DW, Whelan DT. (1996) Maternal uniparental isodisomy of human chromosome 14 associated with a paternal t(13q14q) and precocious puberty. *Eur. J. Hum. Genet.* **4**: 153–159.

van der Put NM, Steegers-Theunissen RPM, Frosst P *et al.* (1995) Mutated methylenetetrahydrofolate reductase as a risk factor for spina bifida. *Lancet* **346**: 1070–1071.

Vawter DE, Gervais KG. (1995) Ethical and policy issues in human fetal transplants. *Cell Transplant.* **4**: 479–482.

Weinberg CR, Moledor E, Baird DD. (1994) Is there a seasonal pattern in risk of early pregnancy loss? *Epidemiology* **5**: 484–489.

Wilkie AOM, Amberger JS, Mckusick VA. (1994) A gene map of congenital malformations. *J. Med. Genet.* **31**: 507–517.

Zinaman MJ, Clegg ED, Brown CC, O'Connor J, Selevan SG. (1996) Estimates of human fertility and pregnancy loss. *Fertil. Steril.* **65**: 503–509.

2

Why study human embryos? The imperfect mouse model

Tom Strachan and Susan Lindsay

1. Why we need to study animal models of early human development

Although there are compelling medical and scientific justifications for studying prenatal human development, our current knowledge of this subject is very limited. For the most part, this is due to difficulty in obtaining appropriate material for study. In some countries, current legislation limits research in this area (Burn and Strachan, 1995), and, clearly, ethical considerations have to be carefully taken into account (see Chapter 4 for one perspective). At a practical level not all early developmental stages can easily be accessed in humans (Burn and Strachan, 1995), and the material that is being studied largely represents three developmental windows. At the preimplantation stage, surplus embryos may be made available through *in vitro* fertilization, and current legislation in many countries permits study of such embryos up to the end of the second week of development; a timepoint which was chosen because it immediately precedes the formation of the primitive streak. Two later stages can be accessible in some countries, if ethical approval and maternal consent have been obtained for studying material retrieved following termination of pregnancy [TOP (Burn and Strachan, 1995)]. First-trimester TOP can permit retrieval of intact embryos usually between the start of the fourth week and the end of the eighth week of development, a period when the majority of organogenesis occurs [see Chapter 3 (O'Rahilly and Muller, 1987)]. Fetal material derived from second-trimester TOP is largely restricted to weeks 12–18 of development, a range which is particularly important in understanding brain development.

As the amount of material that can be retrieved for study via the above routes is rather limited, there has been an understandable preoccupation with studying animal models of early human development, particularly at the embryonic and fetal stages. Animal models can provide a plentiful supply of material for the

Molecular Genetics of Early Human Development, edited by T. Strachan, S. Lindsay and D.I. Wilson.
© 1997 BIOS Scientific Publishers Ltd, Oxford.

study of prenatal development and, unlike in humans, the full range of prenatal developmental stages can be accessed. Although itself a contentious area, the use of animal models is, on a global scale, more widely approved than is the study of human prenatal tissue, particularly embryonic tissue. In addition, animal models have been justified because of the general observation that developmental processes and developmental control genes are very strongly conserved during evolution (see below). Thus, although they are never going to be perfect models of early human development, animal models have been extremely valuable in providing important insights into human development. Even as we become increasingly aware of their limitations (see below), animal models will continue to provide useful information, providing an initial experimental system for understanding human development, even if an approximate one. As such, they enable more focused follow-up studies to be conducted on precious human material.

2. Why the mouse is the favourite model of early human development

If we were to rely, in great measure, on extrapolation from an animal model of early human development, it would be preferable to study an animal whose early development is extremely similar to ours. Great apes are potentially the best candidates, simply because they are so closely related to us: their gene products are essentially identical in sequence to ours, and even in the case of non-coding DNA orthologous sequence comparisons typically reveal greater than 97% sequence identity. However, there are enormous practical and financial difficulties in sustaining large primate breeding colonies for research purposes. Ethical considerations and public sensitivities also play a part: those who feel that animal experimentation is warranted, usually on the grounds of advancing medical research, are almost always more comfortable with experimentation on rather small animals.

Other animal models of development, including non-mammalian models, have been widely used; for example, the chick has been a useful model, primarily because of the accessibility of the developing embryo, which facilitates tissue transplantation experiments. However, such models are clearly less related to humans than placental mammals, and so certain mammalian models have been considered to be particularly useful models of early human development. Of these, mouse and rat models have been particularly widely used. Both are generally amenable to experimentation: breeding colonies can be maintained at relatively low expense, generation times are short (around 3 months for mice) and fecundity is high (an average female mouse will produce four to eight litters with an average litter size of about six to eight pups).

Rats, being considerably bigger than mice, have been more generally amenable to physiological experimentation and have been particularly important in modelling fetal brain development, for example. However, the mouse has been the most popular general model of mammalian, and therefore human, development, and also of human developmental disorders (Darling, 1996). This has happened because the laboratory mouse has been particularly widely used in genetic studies. Decades of classical mouse genetics have enabled construction of a detailed genetic map, and while transgenic technology has been applied to a variety of different animal models, the mouse has been the only mammalian model thus far in which it has been possible to introduce predetermined mutations in the germline via homologous recombination in embryonic stem (ES) cells. This additional gene targeting capacity has been particularly helpful in designing mouse models of human disease (Bedell *et al.*, 1997; Clarke, 1994; Darling, 1996; Erickson, 1996) and in studies of the function of genes, including genes with a presumptive role in development (St. Jacques and McMahon, 1996). Such procedures now encompass a range of sophisticated germline gene manipulations, including, in addition to 'classical' knockouts, the introduction of subtle point mutations (Hasty *et al.*, 1991), inducible knockouts (Kuhn *et al.*, 1995) and chromosome engineering (Ramirez-Solis *et al.*, 1995). The availability of cultured mouse ES cells has also facilitated the construction of transgenics containing whole yeast artificial chromosomes [which can, for example, permit analysis of large human genes or gene clusters (Peterson *et al.*, 1997)] or even very large fragments of human chromosomes (Tomizuka *et al.*, 1997).

3. General limitations of mouse models

Increasingly, we have become aware of the limitations of many mouse models of disease: the phenotypes often diverge considerably from that of the human disorders they were expected to resemble, even in cases where the corresponding phenotypes in humans and mice are known to arise from similar mutations in orthologous genes. In some cases the differences in phenotype are known to be attributable, in part, to variation in the genetic background of the mouse, and it is often important to cross the mutation into different mouse strains in order to assay for phenotypes that most closely resemble the human phenotype (Erickson, 1996). In general, however, interspecific differences in phenotype should not be entirely unexpected. After all, humans and mice are estimated to have diverged from a common ancestor about 80 million years ago, and there are important differences in anatomy, even at the earliest developmental stages (see below), and in some biochemical pathways (Erickson, 1989). In the latter case, a useful example concerns human–mouse differences in the pathways of purine metabolism which are considered to be responsible for the difficulty in modelling Lesch–Nyhan syndrome (Engle *et al.*, 1996; Williamson *et al.*, 1992; Wu and Melton, 1993; Wu *et al.*, 1994; see Bedell *et al.*, 1997 for an overview). We have also become aware of a whole class of human disorders, those which result from unstable expansion of triplet repeats, which have no known natural counterparts in rodents. These disorders can be modelled in mice when genes or even exons with suitably long triplet repeats are artificially inserted into the germline (Mangiarini *et al.*, 1996). Possibly, therefore, unstable triplet repeat expansions could occur naturally in rodents and the failure to detect them is due simply to difficulties in ascertainment. Alternatively, however, there may be species differences in the mechanisms that have resulted in the modest sizes of such triplet repeats in *normal* populations; there would appear to be greater average length repeats in humans compared to even non-human primates (Dijan *et al.*, 1996; Rubinsztein *et al.*, 1995), and the greater lengths in the human genome could predispose towards the observed human disorders. Finally, we have recently become more aware of fundamental species differences at the DNA and gene expression levels, with many significant differences between humans and mice in both the organization of genomic DNA and in the organization and expression of orthologous genes.

3.1 Human–mouse differences in genome and gene organization

Although the sizes of the mouse and human genomes are comparable (3000 Mb), there are clear differences in subchromosomal organization, reflected in the quite different chromosome banding patterns and in both the number and distribution of CpG islands (Antequara and Bird, 1993; Cross *et al.*, 1997). The total number of genes in each genome is expected to be essentially identical (ca. 80 000), but differences within gene families are not uncommon; for example, the haploid mouse genome has two β-globin genes compared to one in the human genome. While we have long been familiar with the moderate conservation of synteny for autosomal loci in humans and mice, we have more recently come to appreciate that even sex linkage is not completely conserved between the two species; for example, the human *CSF2RA* and *ILRA* genes map to the major pseudoautosomal region but do not have X-linked murine orthologues, instead the known murine homologues are autosomal (Disteche, 1995). Similarly, the presence of a Y-linked *DAZ* gene, a candidate male sterility gene in humans, appears to be restricted to primates (Cooke *et al.*, 1996; Reijo *et al.*, 1996). Comparisons of the polypeptide and coding sequences of orthologous genes in human and mice indicate an approximate average level of about 15% sequence divergence in each case (Makalowski *et al.*, 1996; Murphy, 1993), although many developmentally significant genes, are highly conserved (see below). In addition, however, the extent of human–mouse sequence divergence is so great in some cases that it has been difficult and, in a few cases not thus far possible, to identify murine orthologues of some human genes. Examples include the lipoprotein Lp(a) gene (Lawn, 1996), and some genes which map within, or close to, the major pseudoautosomal region (see Meroni *et al.*, 1996 for references).

3.2 General differences in gene expression

In recent years there has been a growing appreciation of differences in expression of orthologous genes in humans and mice. In the case of X-linked genes, for example, there are several examples of human genes which escape X chromosome inactivation unlike their murine orthologues (Disteche, 1995). Human–mouse differences in imprinting have also been recorded: the *MAS* proto-oncogene and the neighbouring *IGF2R* gene are not imprinted, unlike their murine orthologues (Riesewijk *et al.*, 1996). Differences are also becoming apparent in the ways in which orthologous human and rodent genes undergo RNA processing. In some cases, sequence divergence at splice sites has resulted in species-specific RNA splicing patterns, as in the case of the *FGF8/Fgf8* comparison (Gemel *et al.*, 1996). Human–rodent differences in promoter usage have also been noted (e.g. Mukai *et al.*, 1991).

4. The current paradigm of conservation in developmental biology

The human–mouse differences in gene organization and expression considered above are general ones, and the relevant question that now needs to be considered is the extent to which they apply to the period of early development. Before going on to discuss human–mouse differences in development, it is instructive to consider the similarities first, given the widely held view that early vertebrate development has been strongly conserved. Such a concept has a very long history and was founded on anatomical and morphological investigations. The German biologist Karl Ernst von Baer (1792–1876) noted in his records of certain vertebrate embryos which he had preserved in spirit but omitted to label: "I am quite unable to say to what class they belong. They may be lizards, or small birds, or very young mammalia, so complete is the similarity in the mode of formation of the head and trunk in these animals. The extremities are still absent, but even if they existed, in the earliest stage of the development we should learn nothing, because all arise from the same fundamental form". This view, and the related idea that embryonic development in mammals has been highly conserved (see *Figure 1*) are now known to be considerable simplifications (Richardson *et al.*, 1997).

In recent years molecular genetic approaches have solidly supported the strong conservation of the processes of early animal development. A key finding was the observation that certain *Drosophila* sequences were so crucially important in early development that they had recognizable homologues in mammals (McGinnis *et al.*, 1984). It should be noted, however, that, by and large, this finding startled the scientific community. McGinnis and colleagues had primarily wanted to test whether the homeobox sequences that they had found to be conserved in different species of *Drosophila* were also conserved in worms, and were taken aback by unexpected findings. Lawrence (1992) reports McGinnis's recollection of this seminal discovery: 'I put in some vertebrate DNAs (calf thymus, *Xenopus*, human) hoping for some cross-hybridisation, but really thinking of them as negative controls. When I pulled the first blot out, there were obviously some strongly hybridising fragments in the human, frog and calf lanes. I was so excited my hands were shaking, but most were sceptical about the results and I was a bit disappointed'. McGinnis went on to repeat the experiment successfully several times and people shelved their prejudices that invertebrates and vertebrates had to be completely different. Now we know that the homeobox sequences are not unique: a large range of other transcription factor, signalling molecule and receptor sequences which play crucial roles in early development have been similarly conserved during evolution. As the molecular pathways underlying developmental processes in mammals become uncovered, they are often shown to be very similar to pathways in *Drosophila*, as in the example of WNT signalling (see Figure 5 of Kinzler and Vogelstein, 1996). If such processes are essentially conserved at the molecular level between organisms as evolutionarily distant as fruit flies and humans, then it is not unsurprising that early mammalian development has been so widely assumed to be very strongly conserved. Even in the case of developmental pathways which are known not to be conserved between *Drosophila* and mammals, such as sex determination, there is ample evidence that the molecular components are conserved in humans and mice.

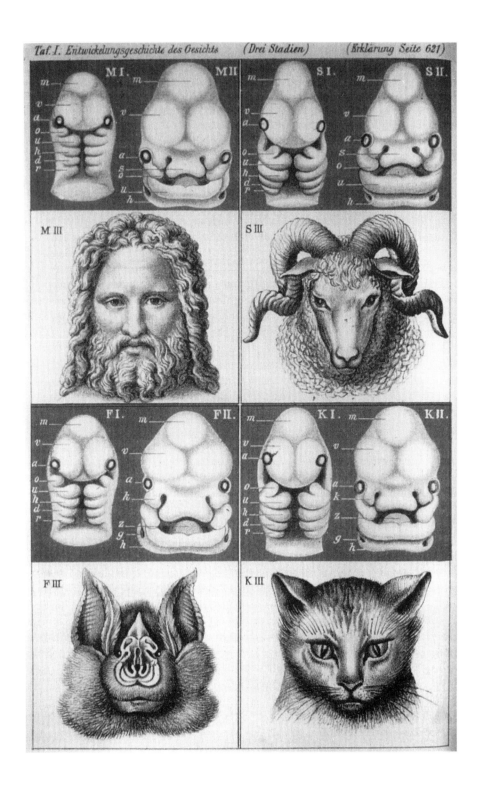

Taf. I. Entwickelungsgeschichte des Gesichts   (Drei Stadien)   (Erklärung Seite 621)

5. Human–mouse differences in early development

The preceding section supports the current paradigm in developmental biology, which emphasizes the conservation of developmental pathways over long periods of evolutionary time. Hopefully, however, this paradigm should not blind us to the increasing evidence for a countercurrent of species differences. At the level of anatomy there are some conspicuous differences between human and mouse embryos (see Chapter 6) and other differences are expected; for example, we have only a very sketchy knowledge of early brain development and the comparative lack of anatomical landmarks in the developing brain have not facilitated species comparisons. Yet, given the enormous differences in the capabilities of the developing brains of mice and men, we should not be surprised to find additional anatomical differences.

Gene expression studies in early development are also identifying species differences. Differences in gene expression in preimplantation embryos are discussed in Chapter 9. In addition, gene expression studies at the postimplantation stage are beginning to identify very significant species differences. Such differences are often being observed in the case of genes which encode essentially identical products in the species being compared (see below). While such findings were not anticipated by many geneticists, early conjectures and prior observations on the evolution of regulatory sequences and coding sequences suggested that species differences in the expression of developmentally significant genes would be significant. In the next section we trace this line of thinking and offer examples from recent work which has revealed species differences in the expression of developmentally significant genes.

6. Species differences in gene expression during early development

The idea that mammalian divergence probably resulted from mutations altering regulatory circuits rather than the sequences of the structural genes was proposed some time ago by various authors, including King and Wilson (1975) and Jacob (1977), and can be traced back to more general considerations on the evolution of phenotypic differences (Zuckerkandl and Pauling, 1965 p.100). More recently, experimental data have been obtained which indicate that species differences in *cis*-acting regulatory sequences are indeed important in altering the expression patterns of functionally identical genes. Transgenic mouse studies have shown that human–mouse differences in *cis*-acting regulatory sequences can explain divergence of expression patterns (reviewed by Cavener, 1992). A useful example

Figure 1. Unlike the adult phenotypes, the gross morphology of early mammalian embryos appears highly conserved. This illustration is reproduced from Haeckel (1874). Panels illustrate the heads of four vertebrates, human (M), sheep (S), bat (F) and cat (K) at three stages in development: the adult form (III), plus two stages in early development (I and II). Note that the drawings of the embryonic forms are now thought to be inaccurate and a considerable simplification (see Richardson *et al.*, 1997).

of general man–mouse divergence revealed by transgenic studies concerns the expression of the elastase inhibitor α_1-antitrypsin. In the mouse α_1-antitrypsin mRNA is abundant in the liver and yolk sac during fetal development, and subsequently in adult liver and kidney. When a human α_1-antitrypsin transgene was expressed in mice, the human gene was seen to be transcribed in the same tissues as observed in the case of mouse α_1-antitrypsin. In addition, however, the human transgene was found to be expressed in a wide variety of tissues in the adult, and in additional tissues in fetal development; investigation of endogenous human α_1-antitrypsin expression confirmed this substantial human–mouse expression difference (Carlson *et al.*, 1988; Kelsey *et al.*, 1987; Koopman *et al.*, 1989).

6.1 Lessons from expression studies of developmentally significant genes in Drosophila

The implication of the above studies of α_1-antitrypsin expression is that the species differences in expression are determined by differences in *cis*-acting regulatory sequences, and similar inferences have been deduced from other such studies in transgenic *Drosophila* (Cavener, 1992). However, the idea that the expression of genes which are critically important in development could also be subject to evolutionarily rapid changes of expression pattern is more novel. Compelling evidence for the importance of sequence divergence at regulatory elements of developmentally critical genes was reported by Li and Noll (1994), who studied the *Drosophila* genes *paired* (*prd*), *gooseberry* (*gsb*) and *gooseberry neuro* (*gsbn*). These genes are related to the vertebrate PAX genes, encoding proteins with similar paired domains and paired-type homeodomains, and are thought to have evolved by ancestral gene duplication. They have distinct spatiotemporal expression patterns in development and distinct developmental functions. Of the three, the *prd* gene is expressed first in the developing embryo, and is an example of a pair-rule gene, one for which mutations affect *alternate* segments of the embryo. It activates the subsequent expression of the segment polarity gene *gsb*. The *gsb* gene, in turn, activates the expression of *gsbn*, which is expressed mainly in the developing nervous system and is thought to have a role in neural development. Despite the quite different natural expression patterns and the different functions of these three genes, the experiments reported by Li and Noll (1994) indicate that the products of all three genes are functionally equivalent *at the cellular level*; that is, if the different gene products are expressed in the same cells they perform the same function. This was artificially achieved by ectopic expression under the control of a heat-shock promoter, or by expression in mutant embryos of hybrid transgenes with *cis*-acting regulatory sequences of one gene driving expression of the coding sequence of another; for example, one set of experiments used a transgene containing the *prd* coding sequence coupled to the *cis*-acting regulatory sequences of the *gsb* gene. The resulting hybrid gene could express a Prd protein in cells where the Gsb product is normally expressed, and when introduced into a mutant *gsb⁻* embryo, the Prd product was able to substitute for the absent Gsb product and rescue the mutant phenotype.

The above experiments suggested that divergence of *cis*-acting regulatory elements could lead not only to divergence of expression pattern but that the resulting difference in spatiotemporal expression pattern could, *by itself,* lead to divergence in gene function. If so, this would appear to be a useful evolutionary strategy to permit acquisition of novel developmental function, and an important question concerns the evolutionary timescale for this to occur. Recent studies in *Drosophila* suggest that significant alteration in expression pattern may occur over a very short time-span. A good example is provided by the *amd* (α-methyldopa hypersensitive) gene, which is essential for *Drosophila* cuticle development. This gene shows broadly con-served expression patterns in the embryos of *D. melanogaster, D. simulans* and *D. vir-ilis,* but an additional strong expression domain is observed in the central nervous system of *D. simulans* embryos only (Wang *et al.,* 1996). *Drosophila simulans* and *D. melanogaster* are sibling species, thought to have diverged from a common ancestor only 2–5 million years ago (whereas *D. virilis* is more distantly related to both), and the 'sudden' appearance in evolution of a new *amd* expression domain in *D. simulans* has occurred even though the *amd* coding sequence is most probably identical to that of the *D. melanogaster amd* gene. The functional significance of the new expres-sion domain of the *amd* gene in *D. simulans* is uncertain, but the important point is that new expression domains can occur rapidly during evolution, even for very strongly conserved genes. In many cases, a new expression domain may not eventu-ally lead to functional divergence, and the new domain may be tolerated, or eventu-ally lost. In some cases, however, the new expression domain can be expected to lead to functional divergence as in the *prd, gsb* and *gsbn* genes (see above).

6.2 Expression studies of developmentally significant mammalian genes

If we extrapolate from the *Drosophila* studies to mammals, we might expect to see significant differences in embryonic and fetal expression patterns for many sets of orthologous genes in humans and rodents, given that they have diverged from a common ancestor about 80 million years ago. Although the study of gene expres-sion in human postimplantation embryos is still very much in its infancy (Burn and Strachan, 1995), already significant expression differences are being found for orthologous genes in human and mouse embryos. Usually, the comparisons reveal extensive similarity of many expression domains. However, temporal expression differences are not uncommon, which is perhaps not so unexpected given that when different organ systems/tissues are considered there can be some variation in assignment of equivalent developmental stages in the two species. More strikingly, differences are also being found in the spatial expression patterns of some ortholo-gous genes in humans and mice. For example, the human calpain-3 gene, muta-tions in which cause the LGMD2A form of limb girdle muscular dystrophy, shows an expression domain in embryonic heart which is not conserved in mouse. In the developing human embryo, expression is initially seen all over the heart but, as the four chambers form, expression becomes confined to the ventricles, and the ven-tricle-restricted heart expression persists through the subsequent embryonic and fetal periods; in the mouse, by contrast, no calpain-3 expression is detectable in the heart from day 9.5. to birth (F. Fougerousse and J.S. Beckmann, pers. comm.).

While the significance of the above differences is still unclear, already one func-tionally important difference has been suggested: the expression of myosin VIIA. Mutations in the myosin VIIA gene result in Usher's syndrome type 1B in

further evidence of human–mouse differences will be obtained as more and more human genes are analysed. Such differences should reinforce the view that we cannot afford to rely so exclusively on extrapolation from the mouse, even though it is a generally very useful and valuable model of early human development.

Acknowledgement

We are grateful to the Wellcome Trust for continued support for our study of gene expression in early human development.

References

Antequara F, Bird A. (1993) Number of CpG islands and genes in human amd mouse. *Proc. Natl Acad. Sci. USA* 40: 454–461.

Bedell MA, Largaespada DA, Jenkins NA, Copeland NG. (1997) Mouse models of human disease. Part II: Recent progress and future directions. *Genes Dev.* 11: 11–43.

Burn J, Strachan T. (1995) Human embryo use in developmental research. *Nature Genetics* 11: 3–6.

Carlson JA, Rogers BB, Sifers RN, Hawkins HK, Finegold MJ, Woo SLC. (1988) Multiple tissues express alpha₁-antitrypsin in transgenic mice and man. *J. Clin. Invest.* 82: 26–36.

Cavener DR. (1992) Transgenic animal studies on the evolution of genetic regulatory circuitries. *Bioessays* 14: 237–244.

Clarke AR. (1994) Murine genetic models of disease. *Curr. Opin. Genet. Dev.* 4: 453–460.

Cooke HJ, Lee M, Kerr S, Ruggiu M. (1996) A murine homologue of the human *DAZ* gene is autosomal and expressed only in male and female gonads. *Hum. Mol. Genet.* 5: 513–516.

Cross SH, Lee M, Clark VH, Craig JM, Bird AP, Bickmore WA. (1997) The chromosomal distribution of CpG islands in the mouse: evidence for genome scrambling in the rodent lineage. *Genomics* 40: 454–461.

Darling S. (1996) Mice as models of human developmental disorders: natural and artificial mutants. *Curr. Opin. Genet. Dev.* 6: 289–294.

Dijan P, Hancock JM, Chana HS. (1996) Codon repeats in genes associated with human diseases: fewer repeats in the genes of non-human primates and nucleotide substitutions concentrated at the sites of reiteration. *Proc. Natl Acad. Sci. USA* 93: 417–421.

Disteche CM. (1995) Escape from X inactivation in human and mouse. *Trends Genet.* 11: 17–22.

El-Amraoui A, Sahly I, Picaud S, Sahel J, Abitbol M, Petit C. (1996) Human usher 1B/mouse *shaker-1*: the retinal phenotype discrepancy explained by the presence/absence of myosin VIIA in the photoreceptor cells. *Hum. Mol. Genet.* 5: 1171–1178.

Engle SJ, Womer DE, Davies PM, Boivin G, Sahota A, Simmonds HA, Stambrook PJ, Tischfield JA. (1996) HPRT-APRT deficient mice are not a model for Lesch–Nyhan syndrome. *Hum. Mol. Genet.* 5: 1607–1610.

Erickson RP. (1989) Why isn't a mouse more like a man? *Trends Genet.* 5: 1–3.

Erickson RP. (1996) Mouse models of human genetic disease: which mouse is more like a man? *Bioessays* 18: 993–998.

Gemel J, Gorry M, Ehrlich GD, MacArthur CA. (1996) Structure and sequence of human *FGF8*. *Genomics* 35: 253–257.

Gibson F, Walsh J, Mburu P, Varela A, Brown KA, Antonio M, Belsel KW, Steel KP, Brown SDM. (1995) A type VIII myosin encoded by the mouse deafness gene *shaker 1*. *Nature* 374: 62–64.

Haeckel E. (1874) *Anthropogenie, oder Entwicklungsgesicht des Menschen.* Engelmann, Leipzig.

Hasty P, Ramirez-Solis R, Krumlauf R, Bradley A. (1991) Introduction of a subtle point mutation into the *Hox-2.6* locus in embryonic stem cells. *Nature* 350: 243–246.

Jacob F. (1977) Evolution and tinkering. *Science* 196: 1161–1166.

Kelsey GD, Povey S, Bygrave AE, Lovell-Badge RH. (1987) Species- and tissue-specific expression of human α_1-antitrypsin in transgenic mice. *Genes Dev.* 1: 161–171.

King MC, Wilson A. (1975) Evolution at two levels in humans and chimpanzee. *Science* 188: 107–116.

Kinzler KW, Vogelstein B. (1996) Lessons from hereditary cancer. *Cell* 87: 159–170.

Koopman P, Povey S, Lovell-Badge RH. (1989) Widespread expression of human α_1-antitrypsin in transgenic mice revealed by *in situ* hybridization. *Genes Dev.* 3: 16–25.

Kuhn R, Schwenk F, Aguet M, Rajewsky K. (1995) Inducible gene targeting in mice. *Science* 269: 1427–1429.

Lawn RM. (1996) How often has Lp(a) evolved? *Clin. Genet.* 49: 167–174.

Lawrence PA. (1992) *The Making of a Fly.* Blackwell Scientific Publishers Ltd, Oxford, pp. 216–218.

Li X, Noll M. (1994) Evolution of distinct developmental functions by three *Drosophila* genes by acquisition of different *cis*-regulatory regions. *Nature* 367: 83–87.

Li QY, Newbury-Ecob RA, Terrett JA *et al.* (1997) Holt–Oram syndrome is caused by mutations in *TBX5*, a member of the *Brachyury (T)* gene family. *Nature Genetics* 15: 21–29.

Makalowski W, Zhang J, Boguski MS. (1996) Comparative analysis of 1196 orthologous mouse and human full-length mRNA and protein sequences. *Genome Res.* 6: 846–857.

Mangiarini L, Sathasivam K, Seller M *et al.* (1996) Exon 1 of the *HD* gene with an expanded CAG repeat is sufficient to cause a progressive neurological phenotype in transgenic mice. *Cell* 87: 493–506.

McGinnis W, Hart CP, Gehring WJ, Ruddle FH. (1984) Molecular cloning and chromosomal mapping of a mouse DNA sequence homologous to homeotic genes of *Drosophila*. *Cell* 38: 675–680.

Meroni G, Franco B, Archidiacono N, Messali S, Andolfi G, Rocchi M, Ballabio A. (1996) Characterization of a cluster of sulfatase genes on Xp22.3 suggests gene duplications in an ancestral pseudoautosomal region. *Hum. Mol. Genet.* 5: 423–431.

Mukai T, Arai Y, Yatsuki H, Joh K, Hori K. (1991) An additional promoter function in the human aldolase gene, but not in rat. *Eur. J. Biochem.* 195: 781–787.

Murphy PM. (1993) Molecular mimicry and the generation of host defense protein diversity. *Cell* 72: 823–826.

O'Rahilly R, Muller F. (1987) *Developmental Stages in Human Embryos.* Carnegie Institute, Washington. Publication 637.

Peterson KR, Clegg CH, Li Q, Stamatoyannopoulos G. (1997) Production of transgenic mice with yeast artificial chromosomes. *Trends Genet.* 13: 61–66.

Ramirez-Solis R, Liu P, Bradley A. (1995) Chromosome engineering in mice. *Nature* 378: 720–724.

Reijo R, Seligman J, Dinulos MB, Jaffe T, Brown LG, Disteche CM, Page DC. (1996) Mouse autosomal homologue of *DAZ*, a candidate male sterility gene in humans, is expressed in male germ cells before and after puberty. *Genomics* 35: 346–352.

Richardson MK, Hanken J, Gooneratne ML, Pieau C, Raynaud A, Selwood L, Wright GM. (1997) There is no highly conserved embryonic stage in the vertebrates: implications for current theories of evolution and development. *Anat. Embryol.* 196: 91–106.

Riesewijk AM, Schepens MT, Mariman EM, Ropers H-H, Kalscheurt VM. (1996) The *MAS* proto-oncogene is not imprinted in humans. *Genomics* 35: 380–382.

Rubinsztein DC, Leggo J, Amos W. (1995) Microsatellites evolve more rapidly in humans than in chimpanzees. *Genomics* 30: 610–612.

St.-Jacques B, McMahon AP. (1996) Early mouse development: lessons from gene targeting. *Curr. Opin. Genet. Dev.* 6: 439–444.

Strachan T, Abitbol M, Davidson D, Beckmann J. (1997) A new dimension for the Human Genome Project. *Nature Genetics* 16: 126–132.

Tomizuka K, Yoshida H, Uejima H *et al.* (1997) Functional expression and germline transmission of a human chromosome fragment in chimaeric mice. *Nature Genetics* 16: 133–143.

Wang D, Marsh JL, Ayala FJ. (1996) Evolutionary changes in the expression pattern of a developmentally essential gene in three *Drosophila* species. *Proc. Natl Acad. Sci. USA* 93: 7103–7107.

Weil D, Blanchard S, Kaplan J *et al.* (1995) Defective myosin VIIA gene responsible for Usher syndrome type 1B. *Nature* 374: 60–61.

Williamson DJ, Hooper ML, Melton DW. (1992) Mouse models of hypoxanthine phosphoribosyltransferase deficiency. *J. Inherit. Metab. Dis.* 15: 665–673.

Wu C-L, Melton DW. (1993) Production of a model for Lesch–Nyhan syndrome in hypoxanthine phosphoribosyltransferase-deficient mice. *Nature Genetics* **3**: 235–240.

Wu X, Wakamiya M, Vaishnav S, Geske C, Montgomery C Jr., Jones P, Bradley A, Caskey CT. (1994) Hyperuricemia and urate nephropathy in urate oxidase-deficient mice. *Proc. Natl Acad. Sci. USA* **91**: 742–746.

Zuckerkandl E, Pauling L. (1965) Evolutionary divergence and convergence in proteins. In: *Evolving Genes and Proteins* (eds V Bryson, HJ Vogel). Academic Press, New York, pp. 97–166.

3

The Carnegie Staging of human embryos: a practical guide

Philip Bullen and David Wilson

1. Historical background

The ordered study of human embryology was pioneered by Wilhelm His (1880–1885) at the end of the last century, advanced by developments in microtome equipment and the introduction of formalin as a fixative. Although His naturally arranged embryos studied in developmental order, it was not until 1914 that Mall divided human embryonic development into stages (Mall, 1914), following a precedent set in the study of other species. It was Mall who founded the Carnegie Institution of Washington's Department of Embryology, and his collection of embryos, commenced in 1887, was the foundation for the Carnegie Collection. His successor, Streeter, subsequently produced a classification of divisions that he termed 'developmental horizons' (Streeter, 1942). By 1951 Streeter had completed a 23-stage system (Streeter, 1951), ending the embryonic period with the appearance of marrow in the humerus at about 56 days. Such a classification is largely the basis of the current definitive Carnegie Staging system, although the latter was not formalized until 1987 by O'Rahilly and Muller (1987). This final text was therefore able to draw upon nearly 100 years of work, including illustrations and photographs, as well as the authors' own extensive assessment of the Carnegie specimens.

2. The Carnegie Stages of human embryo development

Data from the Carnegie collection of 695 fixed human embryos, has led to the division of the 56 day human embryonic period into 23 stages (O'Rahilly and Muller, 1987). The defining features of these stages are summarized in *Table 1* and include the size and age ranges into which members of each stage usually fall. Both external morphology and highly detailed histological information have been determined.

Molecular Genetics of Early Human Development, edited by T. Strachan, S. Lindsay and D.I. Wilson.
© 1997 BIOS Scientific Publishers Ltd, Oxford.

Table 1. The Carnegie Stages of human embryo development

Carnegie Stage	Size (mm)	Age (days)	Features
1		1	Fertilized oocyte
2		2	Morula
3		4	Blastocyst
4		5–6	Blastocyst attaching to endometrium
5		7–12	Implanted blastocyst; bilaminar disc
6		13	Chorionic villi; trilaminar disc with primitive streak
7	0.4	16	Notochordal process
8	1–1.5	18	Notochordal canal from primitive pit
9	1.5–2.5	20	Somites appear (1–3); neural groove; rostral neural folds; cardiac primordium
10	1.5–3	22	Somites 4–12; neural folds fusing centrally; asymmetric heart tube
11	2.5–3.5	24	Somites 13–20; rostral neuropore closing; pharyngeal arches 1 and 2
12	3–5	26	Somites 21–29; caudal neuropore closing; pharyngeal arches 1–3; may have upper limb bud
13	4–6	28	Somites 30+; four limb buds visible; otic vesicle closes off
14	5–7	32	Open lens pit; elongated tapering upper limb; (optic cup, endolymphatic appendage)
15	7–9	33	Lens vesicle closed; nasal pits; hand plate forming; telencephalon distinct; antitragus beginning
16	8–11	37	Retinal pigment visible; pharyngeal arch 3 receding; auricular hillocks appearing; thigh, leg and foot discernible
17	11–14	41	Nasofrontal grooves distinct; auricular hillocks all present; hand plate digital rays; relative head enlargement
18	13–17	44	Hand plate notched; elbow identifiable; tip of nose seen in profile; rudimentary eyelids; specific parts of external ear forming
19	17–20	47	Trunk longer and straighter; head flexion no longer right angle to body; limbs point almost straight forward; prominent toe rays
20	21–23	50	Elbows bent; hands flexed at wrist but not close to meeting in midline; superficial cranial vascular plexus seen as line
21	22–24	52	Fingers longer with swollen distal 'tactile pads'; hands and feet about to meet in midline; head vascular plexus nearer vertex now
22	25–27	54	Head vascular line ¾ way from ear–eye level to vertex; fingers may overlap; tragus/antitragus more substantial; eyelids thickening
23	28–30	56	Limbs longer with more mature proportions; head rounder and vascular plexus nearly at vertex; eyelids may start to fuse at margins

Based largely on data from O'Rahilly and Muller.

the following Carnegie stages and *Table 2* is an interpretation of the practical land-marks of unfixed embryos which we have found most useful for reliable and rapid staging. *Figures 1–14* are examples of embryos between stages 8 and 23 (except stage 9). The images were obtained from a Zeiss SV11 Stereomicroscope and captured using an Imaging Associates KS200 video capture system. The images illustrate specific landmarks and may assist staging; they do not define each stage with 'diag-nostic images' by illustrating all the stage-specific landmarks.

5. Conclusion

The timely establishment of an international standard for human embryo staging in 1987 is likely to be of increasing importance and significance in the next few years, as the new investigative tools of molecular genetics are applied to explo-ration of human development. We realize that species such as the mouse and rat have provided much valuable information in developmental biology and will undoubtedly continue to be extremely useful, however, the recognition of devel-opmental differences between mammalian species and the limitations of animal models of human disease will inevitably direct comparison between these species and humans. A fundamental requirement for embryonic comparison of different species is a definition of developmental age. Whilst one may anticipate in the future that molecular markers indicative of specific developmental stages will be identified, which will facilitate interspecies comparison, we have had to practise an anatomical approach which is essentially the same as that practised by other developmental biologists.

Acknowledgements

The colour images of the embryos have been included with the generous support of Carl Zeiss and Imaging Associates and we are grateful for their help.

References

His W. (1880–1885) *Anatomie menschlicher Embryonen*. Vogel, Leipzig.

Mall FP. (1914) On stages in the development of human embryos from 2 to 25 mm long. *Anat. Anz.* 46: 78–84.

O'Rahilly R, Muller F. (1987) Developmental stages in human embryos. *Carnegie Institution of Washington. Publication No. 637.*

Penttila A, McDowell E, Trump BF. (1975) Effects of fixation and post-fixation treatments on volume of injured cells. *J. Histochem. Cytochem.* 22: 251–270.

Streeter GL. (1942) Developmental horizons in human embryos. Description of age group XI, 13 to 20 somites, and age group XII, 21 to 29 somites. *Carnegie Institution of Washington. Publication No. 541, Contrib. Embryol.* 30: 211–245.

Streeter GL. (1951) Developmental horizons in human embryos. Description of age groups XIX, XX, XXI, XXII and XXIII; the fifth issue of a survey of the Carnegie collection. *Carnegie Institution of Washington. Publication No. 592, Contrib. Embryol.* 34: 165–196.

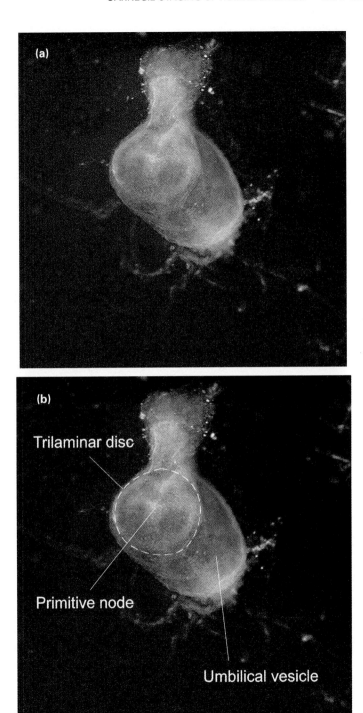

Figure 1. Stage 7/8 human embryo: 16–18 days. (a, b) Dorsal view of a trilaminar disc; the primitive node and the umbilical vesicle can be seen. (c, d) A lateral view of the same embryo showing a collapsed amniotic cavity over the dorsal surface of the embryo. The umbilical vesicle has remained expanded. The distinction between stage 7 and stage 8 is defined histologically.

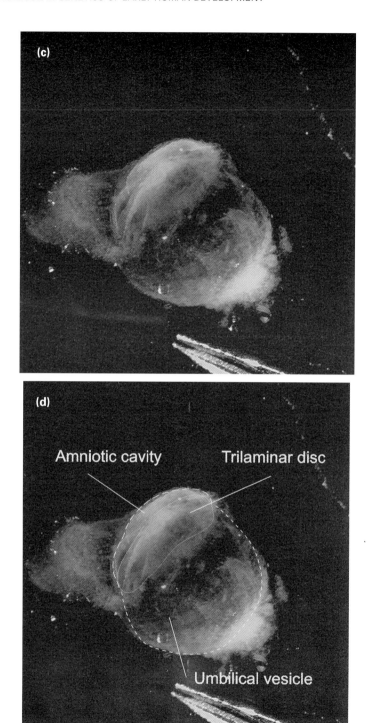

Figure 1. Continued.

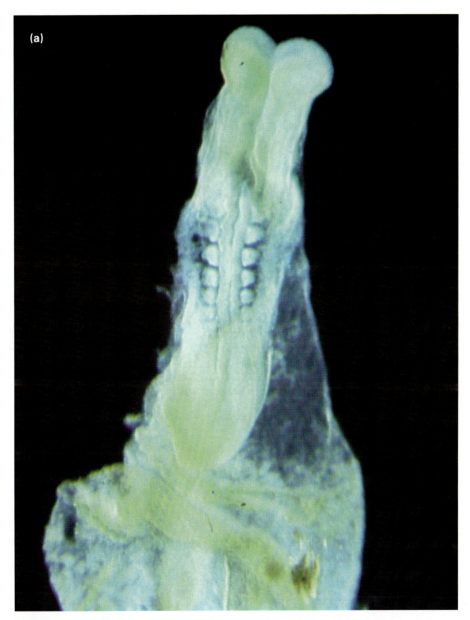

Figure 2. Stage 10 human embryo: 22 days. (a) Dorsal view of an embryo with 7 somite pairs; a disrupted yolk sac is attached on the bottom of the image. The neural tube has fused at the level of the somites but remains open in regions of the hindbrain and forebrain. (b) Ventral view of the cranial end of the same embryo. The opening into the primitive foregut (fg) is clearly visible and the primitive heart tube (ht).

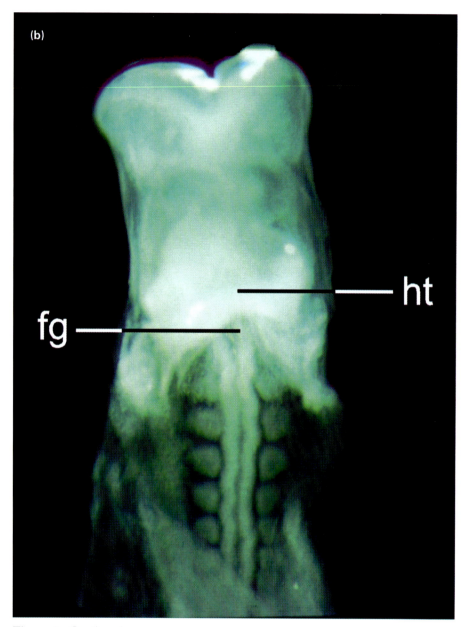

Figure 2. Continued.

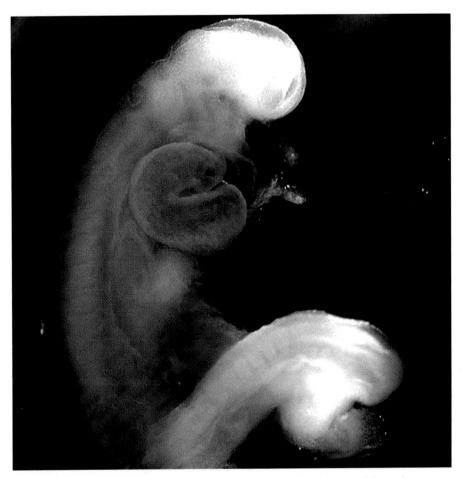

Figure 3. Stage 11 human embryo: 24 days. Pharyngeal arches 1 and 2 can be seen. The anterior neural folds have fused but the rostral neuropore remains open; the primitive heart tube has looped to the right. As is common for stages 11–13, the axis rotation seen in this embryo makes counting somites difficult.

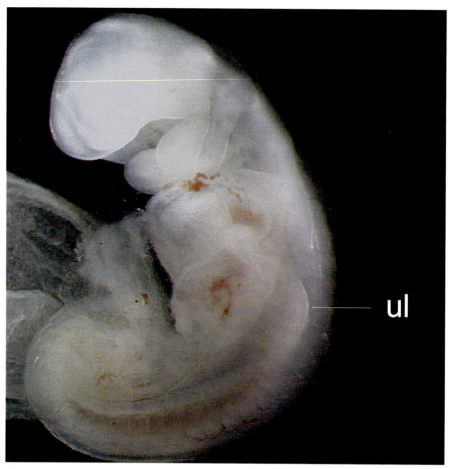

Figure 4. Stage 12 human embryo: 26 days. The upper limb buds (ul) are present but the lower limb buds are absent. In addition the pharyngeal arches, 1–3, are visible and the rostral neuropore has closed.

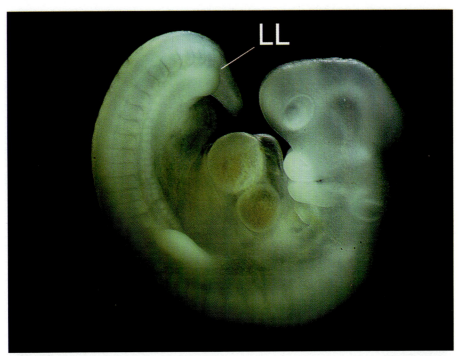

Figure 5. Stage 13 human embryo: 28 days. The upper and lower limb (LL) buds are present. In addition the atrioventricular sulcus can be seen which marks the division between the cardiac atria and ventricles.

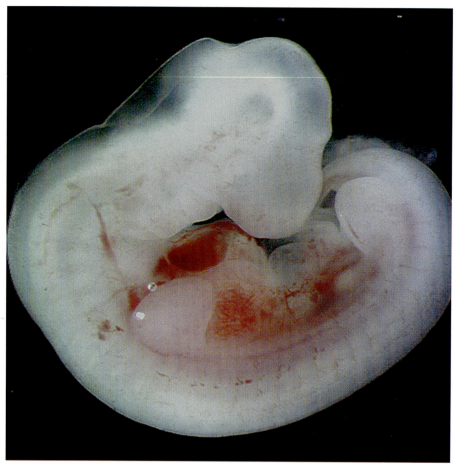

Figure 6. Stage 14 human embryo: 32 days. The upper limb bud is tapering and the optic pit is open. The embryo shown is unfixed; it is much easier to visualize the optic pit following fixation, particularly with Bouin's fixative, when the tissues are less translucent.

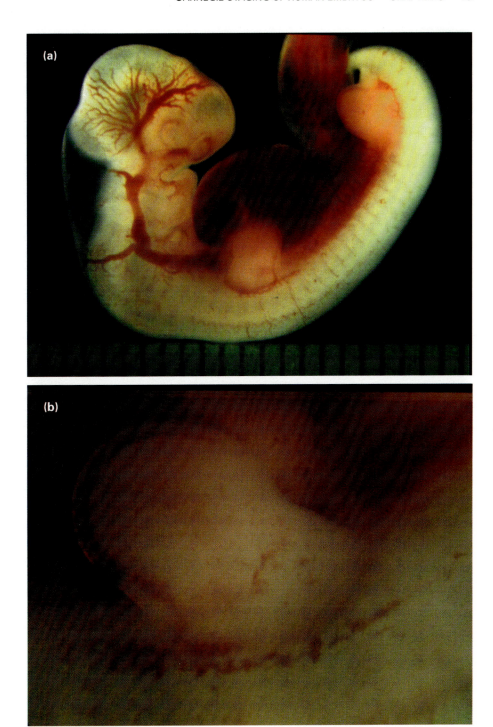

Figure 7. Stage 15 human embryo: 33 days. (a) The nasal pits, telencephalon and hand plates can be seen but it can be difficult to identify closure of the lens pit in unfixed embryos. (b) The upper limb broadens distally to form an early hand plate.

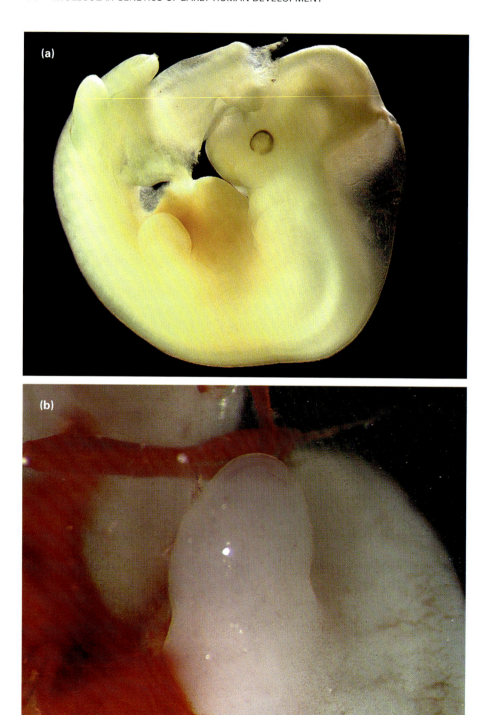

Figure 8. Stage 16 human embryo: 37 days. The lower limb has a foot plate (b) and retinal pigment is clearly seen.

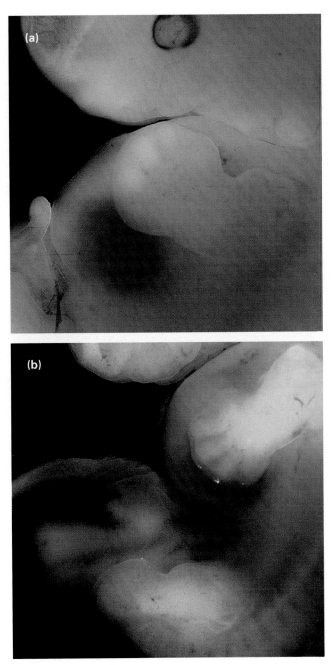

Figure 9. Stage 17 human embryo: 41 days. The nasofrontal groves can be identified and hand plate digital rays. A foot plate can be seen (b) but no digital rays.

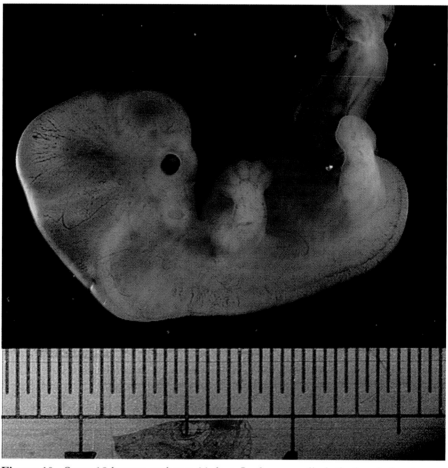

Figure 10. Stage 18 human embryo: 44 days. In the upper limb the hand plate is notched and an elbow is beginning to be identifiable.

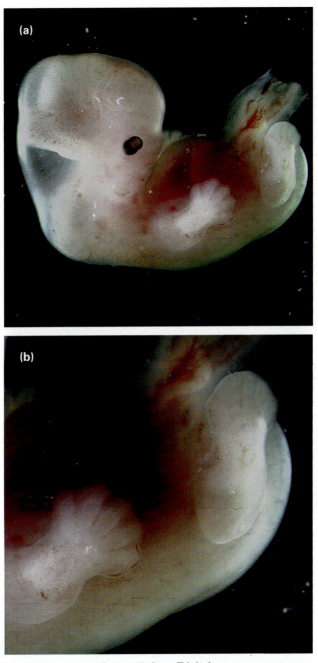

Figure 11. Stage 19 human embryo: 47 days. Digital toe rays are more prominent (b) and do not have notches but features such as head deflexion and limb straightening can be subjective. Stages 18 and 19 are very similar and in practice we have found them two of the most difficult stages to distinguish and discrimination is best made at the time of microscopic examination. The distinction between the embryos shown in *Figures 10* and *11* has largely been made on limb morphology rather than deflexion of the head.

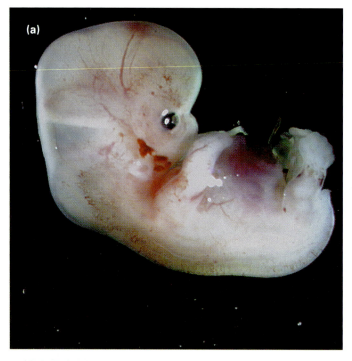

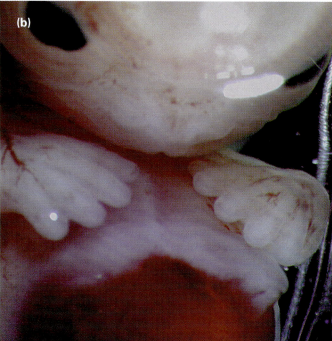

Figure 12. Stage 20 human embryo: 50 days. The toes are notched but the fingers are now stubby digits rather than merely notched (b). The elbows are bent and the hands are held apart. A vascular plexus has formed on the lower lateral region of the head (see also *Figure 14*).

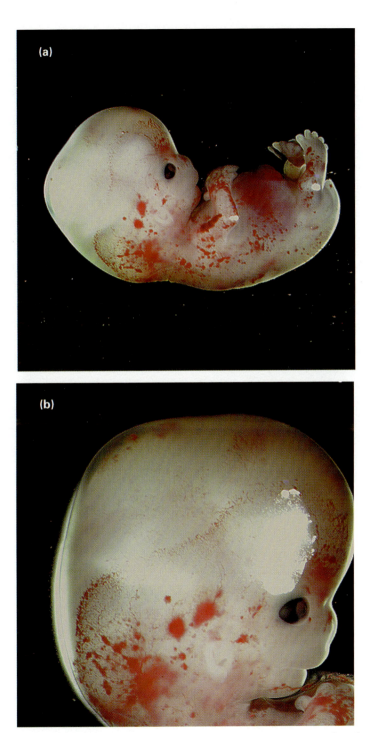

Figure 13. Stage 21 human embryo: 52 days. The toes are separate and no longer just notches. The vascular plexus on the head has moved toward the vertex (see also *Figure 14*).

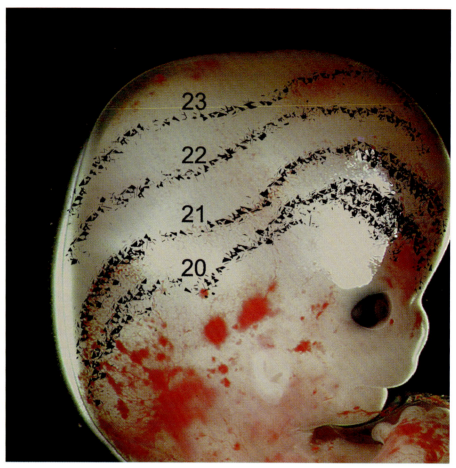

Figure 14. Stages 20–23 human embryo: 50–56 days. An illustration of the vascular plexus at different Carnegie Stages 20–23.

4

The ethics of human embryo studies

Rebecca Bennett and John Harris

1. Introduction

No discussion of the ethics of human embryo research can avoid the question of the moral status of the human embryo, for it is the moral status of the individual which determines the ethics of how it is to be treated, which settles the question of its rights. Human conduct in all societies presupposes that different individuals have different moral status or, to put the same point in different terminology, human conduct presupposes that different sorts of lives have different *value*.

Even if we are vegetarians, our daily food will involve the premature death of some living things, and vegetarians usually accept priorities of importance between different animals and, indeed, between human individuals at different stages of development. Should we prioritize some people over others? If the school is on fire, should we attempt to rescue the children before the teachers? Should the headmaster's 90-year-old granny, who happens to be visiting, be rescued before or after the youngest kindergarten child? What priority should be given to the school cat and the mice in the biology laboratory? While this example may seem fanciful, daily decisions in every society involve these or related questions. How, for example, should we prioritize resources for health care, or, in the case of our present concerns, how should we think of the ethics of treating embryos?

One puzzle, to which we will return, involves the question of how abortion, for example, can be justifiable but not the use of aborted tissue or the use of an embryo destined for a similar fate. A pertinent question must be: if it can be ethical to end the life of an embryo, can it be ethical to allow tissue so yielded to 'go to waste' when it might be used to benefit others?

The view taken by the present authors is based on a particular view of personhood. We cannot attempt a complete account here of the point at which an individual becomes a person, but we can sketch the lines of such an account. Most current accounts of the criteria for personhood follow John Locke in identifying self-consciousness together with fairly rudimentary intelligence as the most important features. The account we believe to be the best uses these (Harris,

Molecular Genetics of Early Human Development, edited by T. Strachan, S. Lindsay and D.I. Wilson.

1985), but argues that they are important *because* they permit the individual to value their own existence. The important feature of this account of what it takes to be a person, namely *that a person is a creature capable of valuing their own existence*, is that it also makes plausible an explanation of the nature of the wrong done to such a being when, for example, it is deprived of existence. Persons who want to live are wronged by being killed because they are thereby deprived of something they value. Persons who do not want to live are not, on this account, harmed by having their wish to die granted, through voluntary euthanasia for example. Non-persons or potential persons cannot be wronged in this way because death does not deprive them of anything they can value. If they cannot wish to live, they cannot have that wish frustrated by being killed. Creatures other than persons can, of course, be harmed in other ways, by being caused gratuitous suffering for example, but not by being painlessly killed, and, as we shall shortly see, the morality of ending their lives will have to take account of the wishes and interests of others.

The life cycle of a given individual passes through a number of stages of different moral significance. The individual can be said to have come into existence when the egg is first differentiated or the sperm that will fertilize that egg is first formed (Harris, 1992). This individual will gradually move from being a potential or a 'pre-person' into an actual person when *they* become capable of valuing *their* own existence, and if, eventually, the individual permanently loses this capacity (for example, those in Persistent Vegetative State), *they* will, on the theory of personhood developed here, have ceased to be a person.

The harm you do in taking a life is the harm of depriving someone of something that they can value, but you may also wrong those who care about them and who value their existence.

The distinction between persons and other sorts of creatures explains the difference between abortion, infanticide and murder and allows us to account for how we might have benefited persons by having saved the lives of the human potential persons they once were, but at the same time shows why we do not harm the potential person by ending that life, whether it be the life of an unfertilized egg or an embryo (Harris, 1985).

This account, very starkly presented, yields a difference in the morality of ending the lives of persons and that of ending the lives of all other creatures including human non-persons.

We can now see how this account gives a clear set of answers to questions about the ethics of human embryo research. Human embryos are not persons; so long as the research culminates in their death prior to becoming persons, no wrong is done to the embryo because it is deprived of nothing it can value, provided, of course, that the embryo does not suffer pain. This proviso is crucial, for ending its life does not exhaust the wrongs that might be done to a living creature that is not a person, including such creatures which are human.

At no stage of its development through to the end of the third trimester and some way beyond does the developing human individual acquire those features that make its moral significance at all comparable to that of normal adult human beings. At no stage does the embryo or the fetus become a creature which possesses capacities or characteristics different in any morally significant way from other animals.

We hope that it would seem wrong to stand in the way of the benefits that might be gained through embryo research unless there are compelling moral reasons to do so. While it appears that the moral status of the embryo is not a compelling reason to prohibit the use of human embryos in research, does the conclusion that the human embryo possess a moral status comparable to cats and canaries allow us to assume that we may put embryos to unlimited use, that we should be allowed to do *anything* to them?

Before we attempt to answer this question, it is important that we clarify just what, in the context of this discussion, we are referring to when we talk about embryo and fetal experimentation. What do we mean by the term 'experimentation'? 'Experiment' is defined in the Shorter Oxford English Dictionary as 'an action or operation undertaken in order to discover something unknown, to test a hypothesis, or establish or illustrate some known truth'. This is clearly a very wide definition of the term, which could easily lead to the classification of a wide variety of activities as embryo or fetal experimentation, including the administering of Zidovudine to fetuses in an attempt to minimize the risk of vertical HIV transmission (Connor *et al.*, 1994) and perhaps the playing of Mozart to the fetus in the womb to enhance musical appreciation. However, in this discussion we will be focusing only on issues raised by research on living embryos and fetuses which exist outside of the womb. There are a number of reasons for this.

1.1 In vivo *experimentation*

Experimentation on live embryos and fetuses *in vivo* introduces a number of further complexities to the problem of experimenting on the unborn, which are specific to this particular situation. Unfortunately, we have insufficient space here to discuss in detail the issues surrounding *in vivo* embryo and fetal experimentation. Nevertheless, we would like to comment briefly on this distinct area.

We would like to emphasize a crucial distinction between, on the one hand, the prospect of research or experimentation upon persons which is undertaken at a pre-personal stage, and, on the other hand, experiments upon embryonic or fetal 'non-persons'. By 'non-persons' here we mean any individual who either cannot become a person or who will not as a matter of fact (perhaps because of an irrevocable decision) become a person.

On the view of personhood we have advanced, while it is not wrong painlessly to experiment upon pre-persons and subsequently end their lives, it is wrong to harm or damage the persons these pre-persons will become. Therefore no experimentation on pre-persons which would be likely significantly to damage the persons that they would become would be ethical.

This has important consequences, of course, for the distinction between experimentation or research on embryos *in vivo*, on the one hand, and *in vitro* on the other. While it is always possible to ensure that *in vitro* embryos are destroyed after experimentation, it is not possible to guarantee this with *in vivo* experimentation. The reasons for this are worth exploring a little further. We take the view, although we will not argue for it here, that it would be genuinely wicked to contemplate forcing a woman to abort a child she wishes to bring to term. Therefore, however solemnly and clearly it might be agreed in advance that experiments on *in vivo* embryos or fetuses would be followed by termination, such decisions are, in principle, almost always vulnerable to a change of mind by the mother. The only exception being where the experimentation is part of the operation that brings about a termination and the mother is unconscious from the time of giving her consent to the completion of the process. Except in these circumstances, however, it could not be guaranteed that a change of mind by the mother would not lead to the survival of a fetus that might have been, damagingly, experimented upon. For these reasons we shall concentrate on the clearer issue of *in vitro* experimentation.

Here again we should not be misled by the confusing aura of associations we have with the term 'experimentation'. It is generally thought that experiments involve serious, damaging and probably painful interventions, whereas an experiment may involve observation only. Procedures which are, for example, well established in animal models and which there is every reason to believe are both highly beneficial and without either serious side-effects or any likely adverse consequences, are also 'experimental' when we move from animals to humans. Suppose a cure for a single gene disorder like sickle-cell disease were discovered which involved deletion of the affected gene and experimenters were highly confident that they could target the gene accurately and that deleting it could be achieved without damage or any likely harmful effects. We might well regard using such an *in vivo* procedure on human embryos who were likely to be adversely affected by the disease as highly desirable, even though it would constitute 'an experiment' on the person into which the pre-person would develop. It would be well worth a small risk because the alternative high probability of contracting the disease would act as the default scenario. Nothing we say in this paper implies that we think such an 'experiment' would be unethical.

2. Dead embryos

For the sake of clarity, it is important to point out that the issues we address in this discussion refer to research on *live* embryos and fetuses outside the womb, and not *dead* fetuses and embryos either acquired by abortion (spontaneous or elective) or which have died or are killed *in vitro*. The use of fetuses and embryos for research that are already dead does not seem to raise any unique ethical issues which need special consideration. If the consent of the mother of the embryo is forthcoming and other things are equal, then it seems that the use of dead fetuses and embryos for research is entirely parallel with other cadaver uses; organ retrieval from deceased children, for example.

With our area of discussion thus focused on research using live embryos and fetuses which exist outside the womb, we will now consider our earlier question of whether the status of the embryo/fetus as a entails that it is justified to put them to unlimited use in the name of research.

3. Research and spare embryos

Some people draw a distinction between so-called 'research' embryos and so-called 'spare' embryos. Spare embryos are those produced in the course of *in-vitro*

fertilization (IVF) treatment but which are not needed for immediate implantation. So-called 'research embryos', on the other hand, are either spare embryos
which have been specifically donated for research by their genetic parents, or
embryos produced from eggs which have been collected as the by-product of IVF
or prior to a sterilization procedure and which have been donated for fertilization
as 'research embryos'. In either case, our view is that so long as the embryo could,
in principle, be implanted, its moral status is the same as that of any potentially
viable embryo.

4. Analogy with abortion

It is our suggestion that there are no grounds for distinguishing the criteria
appropriate for the legitimacy of research on embryos from the criteria for legiti-
mate abortion and that embryo experimentation should be permissible on the
same terms as abortion. A society which holds that abortion is acceptable to preserve and promote the health and well-being of its members, or some of them
(namely mothers), should surely take the same view of research on embryos for
the same reasons.

Yet it has been suggested that the use of embryos/fetuses on the same terms as
abortion should not be permitted; for instance, it was the Warnock Committee's
suggestion that:

'It shall be a criminal offence to handle or use as a research subject any live
human embryo derived from *in vitro* fertilization beyond that limit (i.e. four-
teen days after fertilization)'

(Warnock, 1985)

Is there any good reason why embryo/fetus experimentation should not be permissible on the same terms as permissible abortion? Is it justifiable to allow abortion up to 24 weeks of gestation but reject the same time limit for research on
embryos created *in vitro* (Harris, 1983)?

It would seem that, whatever the view one takes of the moral status of the unborn
human being, any view which holds abortion to be permissible, however limited the
circumstances of this permission, must do so on the basis that in certain circumstances the fetus may have its interests 'trumped' by the rights or interests of the
mother. Thus, where abortion is permitted, it is usually undertaken in order to benefit the mother in some sense, and this is permissible because the rights and interests of the embryo/fetus are subordinate to those of the woman carrying it. If the
view of personhood advanced here is persuasive, we have a justification for this pri-
oritization of the mother's interests, that justification is simply that the fetus is not
a person and its mother is. It may be, of course, that the only benefit for the mother
that is held to be significant enough to outweigh the fetus's claim to life is the benefit of saving her life. It would surely be incongruous not to allow experimentation
on embryos and fetuses on the same terms as abortion. (This might mean, for example, that in societies that only allow abortion to save the lives of pregnant women,
embryo experimentation would only be allowed in comparable cases, that is, where
it is highly likely that the results of this research would save lives.)

5. Pain and suffering

It does not follow from our suggestion that killing an embryo does no wrong to the embryo, that there is nothing morally wrong with doing other things to or with it; for example, we may consistently hold that it is wrong to inflict suffering on creatures whom it is permissible to kill, and that if they are to be killed it must be done without significant pain. This is, we suppose, the attitude of most people to the killing of animals for food and other human purposes. Where pain is inflicted on such creatures (and where it cannot be claimed that it is for the creature's own good as it would be, for example, in surgical operations to save the lives of animals) the gains for humanity must be of an importance to warrant the pain to the creatures involved.

If, as seems likely, the embryo is not capable of feeling pain in the first few weeks of life, because it lacks a sufficiently established nervous system, then this reason for not interfering with it cannot apply, and it will not apply, for similar reasons, where adequate anaesthesia is used. We should also note, and be concerned, that terminations of pregnancy are sometimes carried out very late and are performed in circumstances that are careless of the suffering that might be inflicted on the fetus.

So, although the gratuitous infliction of pain gives us one reason to object to experiments on the embryo, it is an objection that can be very easily met (Harris, 1985).

To summarize the argument so far, we have suggested that if it is accepted that the human embryo and fetus are not as yet persons, a compelling argument can be presented that the destruction of these embryos/fetuses is not morally wrong as long as there is no significant suffering caused thereby (by no significant suffering we mean the level of suffering sentient, conscious individuals could be reasonably expected to bear). If, as a society, we accept that abortion should be permissible up to the 24th week of gestation then there would seem no compelling reason why the same terms should not, in principle, govern the use of embryos and fetuses for experimentation.

6. The slippery slope?

The claim that, as pre-persons, embryos and fetuses have no intrinsic moral value and are not harmed by their destruction would seem to support a permissive approach to both abortion and experimentation on embryos and fetuses.

If we accept that human individuals only attain a 'right to life', and thus protection from destruction, by the possession of 'personhood' as described above, does this approach commit us to the conclusion that embryos and fetuses may be used for *any* research purpose, however trivial, simply on the grounds that as pre-persons there is no harm in their destruction? There are a number of reasons why this is not so.

6.1 Pre-persons, non-persons and ex-persons: Anything goes?

If we accept that destruction of fetuses and embryos is justifiable and we also accept that neonates can also be said to be pre-persons, is it justifiable to make similar claims about their destruction? Intuitively, many would claim that, if embryos and fetuses are to be experimented on, this experimentation should be restricted to embryos and fetuses under 24 weeks, suffering should be kept to a minimum and only 'worthwhile' objectives and information that could not be gleaned in any alternative way should be sought by these methods (we have not of course forgotten that we are talking overwhelmingly about *in vitro* experimentation and that 24 weeks is currently rather beyond present abilities to sustain live embryos outside the womb, however, this clearly may not always be the case). Does the personhood view of moral status reveal that such views are irrational intuitions or are there convincing and cogent reasons which complement the personhood view that would support upholding these intuitive statements?

It would seem that there are a number of powerful reasons why, while accepting that the destruction of embryos or fetuses is not morally wrong, it does not follow that it is desirable to move from this premise to the conclusion that pre-persons may be destroyed for *any* reason, however trivial, and that permitting abortion and embryo experimentation will necessarily lead to infanticide and experimentation on infants.

The personhood view provides an account of the rights and interests of the individual, an account which turns on the combination of harms and wrongs that the individual can suffer. From this perspective there is nothing wrong or indeed harmful about killing a non-person painlessly, or experimenting on it painlessly and then killing it. However, it is important to make it clear that this personhood view, while useful, does not exhaust the ethics of ending the lives of living individuals.

Although as potential or pre-persons, embryos, fetuses and neonates can, on the personhood view, be said not to have intrinsic value, this does not necessarily mean they are valueless and that their destruction should always be permissible. It is clear that human pre-persons may have great extrinsic value; so that, as Mary Anne Warren suggests, the:

> 'needless destruction of a viable infant inevitably deprives some person or persons of a source of great pleasure and satisfaction, perhaps severely impoverishing their lives.'
>
> (Warren, 1982)

While the embryo, fetus or neonate cannot be wronged by being killed because death does not deprive them of something they can value, their destruction may wrong those who do value them.

Non-persons may also have significant symbolic value, and this should not be ignored. Bonnie Steinbock, for example, argues that just because entities do not have significant moral status this does not mean that it does not matter how we treat them:

> 'Like human corpses, human fetuses are human. Like trees, they are alive. Like flags, they have, for many people symbolic significance. All of these features may give rise to moral reasons for treating or not treating them in certain ways.'
>
> (Steinbock, 1992)

doing something good or useful is necessarily a moral improvement on doing nothing good or useful. Our suggestion that the degree of concern over the wisdom of preserving frozen embryos for more than 5 years was somewhat exaggerated, has been, if anything, confirmed by the *Human Fertilisation and Embryology (Statutory Storage Period for Embryos) Regulations 1996*. In the wake of the furore over the so-called 'orphaned embryo opera', which resulted in the destruction of around 4000 embryos, these 1996 regulations extended the permitted storage period from 5 to 10 years. This doubling of the time permitted for storage, without any significant new evidence being adduced as to safety margins, clearly demonstrates that only relatively tentative fears about the dangers of long-term preservation were sufficient to justify mass destruction of embryos.

We live in a society which accepts the inevitability of the creation of many thousands of embryos which are not destined, for one reason or another, to be implanted. We have existing legislation which provides for their use either for research or, failing that, their eventual destruction. Our view is that it cannot be *more* ethical simply to destroy the embryos without putting them to any significant beneficial use than it would be to put them to such use prior to their destruction.

When and if it becomes possible to extend the age to which embryos can be cultured, or to contract the point of viability for a premature fetus, we will have to face the question of whether, if it is permissible to destroy such embryos or fetuses, it is not also permissible to study them or experiment upon them prior to destruction. Those who find the idea of experimenting on embryos beyond 14 days disturbing will need a way of distinguishing between the moral status of the embryo before 14 days and its moral status subsequent to that. This distinction will have to make sense of the legitimacy of *destroying* the fetus after 14 days and explain why while it may be destroyed it cannot be used for painless beneficial experiments or observations prior to destruction. Doubtless there are many such distinctions that would serve this purpose but so far they have evaded the diligent investigations of the present authors. Of course we as a society can simply *decide* to draw this distinction at 14 days without any plausible rationale. The problem is, of course, that if drawing the distinction at this point costs nothing or little in terms of lost beneficial research, then there is no reason not to draw it at a point at which, however irrationally, people feel comfortable with the distinction. However, once there are good reasons in terms of likely beneficial research for moving the dividing line, then the lack of a plausible rationale for its prior position makes pressure for relaxation morally irresistible. Worse, it becomes difficult to set a new limit that can command respect.

So far as we are aware, only one argument for limiting research on embryos to embryos up to 14 days development has ever been produced. It was contained in the Warnock Report (Warnock, 1985) and enshrined in subsequent United Kingdom legislation. The only positive reason contained in the Warnock Report for setting the permissible limits to research at 14 days is that this is (roughly) the point at which the so-called 'primitive streak' appears, which is the first sign of the development of a central nervous system in the embryo. The rationale given by Warnock for the importance of this landmark is not, interestingly, that it is evidence of some important change in the moral status of the embryo, but rather that it is 'the latest stage at which identical twins occur', and, therefore, the first point at which it is clear that the embryo could become a single individual or more than a single individual. Why this point is of *moral* relevance has never been explained (Harris, 1985). Warnock's other great concern, that: 'some precise decision must be taken in order to allay public anxiety' is well taken, but it indicates no particular point. This chapter has suggested a point which does make some sense and which can be applied consistently with other accepted landmarks in relation to the moral importance of the embryo or fetus. We have suggested that the justifications for, and the legal limits upon, abortion provide an appropriate framework for experiments upon *in vitro* embryos, where there is no question of the *in vitro* embryo being subsequently implanted and becoming viable. The principle, already accepted by society in the case of abortion, would be that the interests of embryos are subordinate to the health and welfare of persons properly so-called.

We have tried to outline a morally and scientifically plausible distinction between creatures, persons, which are justifiably protected against experimentation and those which are not. We are not, of course, advocating experimentation on embryos beyond 14 days; no such experiments should be advocated without strong scientific reasons for so doing (in terms of the significance of the knowledge that might thereby be gained) and without identifying the important expected benefits for humankind (in terms of the applicability of the knowledge to the solution of problems affecting the health or welfare of people). We are, however, suggesting that those who would or may have reason to advocate such experiments for good and powerful reasons are entitled to a commensurate response detailing the good and powerful objections. Objections to experiments which would likely result in the alleviation of human misery and which might restore health or prevent disease, would have to point to something stronger than the fact that people found the idea of such research disquieting or disturbing.

References

Connor EM, Sperling RS, Gelber R et al. for the Pediatric AIDS Clinical Trials Group. (1994) Reduction of Maternal–infant transmission of human immunodeficiency virus type 1 with zidovudine treatment. *N. Engl. J. Med.* **311**: 1173–1180.

Harris J. (1983) *In vitro* fertilisation: The ethical issues. *The Philosophical Quarterly* **33**: (No. 132) 217–238.

Harris J. (1985) *The Value of Life: An Introduction to Medical Ethics*. Routledge & Kegan Paul, London.

Harris J. (1992) *Wonderwoman and Superman*. Oxford University Press, Oxford.

Steinbock B. (1992) *Life Before Birth*. Oxford University Press, New York.

Warnock M. (1985) *A Question of Life — The Warnock Report on Human Fertilization and Embryology*. Basil Blackwell Publishers, Oxford.

Warren MA. (1984) On the moral and legal status of abortion. In: *The Problem of Abortion*, 2nd edn (ed. J Feinberg). Wadsworth Publishing Company.

5

Establishing new human embryo collections for documenting gene expression in early human development

Philip Bullen, Stephen C. Robson and Tom Strachan

1. Introduction

Our knowledge of early human development is very limited, largely because the potential for direct study of the human embryo is restricted by both practical and ethical considerations. Yet there is a compelling need to study human embryos, which is justified on both scientific and medical grounds. We need to be able to understand more about normal human development in order to appreciate fully unique human functions and we have come to realize that, although animal models of early human development can be very valuable, they may also be misleading and are at best only approximations of the human system (see Chapter 2). We also need to be able to understand abnormal human development, given that 2–3% of live births have congenital abnormalities, which can often be severe. Archived human embryo collections have been very useful guides to understanding human anatomy in early development, but we now need new human embryo collections in order to document gene expression in early human development. In this chapter we describe the general procedures which have been, and are being, used to collect and store human postimplantation embryos, and the ways in which relevant data are being recorded. We give a brief summary of archived human embryo collections and outline the need for establishing new human embryo collections. Finally, we present our experiences in establishing and curating the Newcastle Human Embryo Collection.

2. Sources of embryonic material for research

The study of early human development can be conducted at two levels:

Molecular Genetics of Early Human Development, edited by T. Strachan, S. Lindsay and D.I. Wilson.
© 1997 BIOS Scientific Publishers Ltd, Oxford.

(i) *In vivo* by the study of embryos and fetuses *in utero*;
(ii) *Ex vivo*, either by fertilizing oocytes *in vitro* and culturing the resulting preimplantation embryos, or by accessing embryonic or fetal material as a result of spontaneous abortion or termination of pregnancy.

In the last 15 years *in vitro* fertilization (IVF) technology and practice has increased enormously. Research on surplus embryos has received increasingly close scrutiny, particularly as embryo replacement during treatment is normally performed after only 24–48 h. In the UK such pre-embryos can be cultured for up to 14 days before they must be destroyed; a date that was selected because it immediately precedes the formation of the primitive streak, the first indicator of the lateral symmetry that characterizes the embryo proper. Gene expression and cytogenetic studies (Plachot *et al.*, 1987) have been performed on these preimplantation embryos yielding valuable information on human development up to this stage (see Chapters 8 and 9). At the postimplantation level, human embryos have been retrieved following spontaneous abortion or termination of pregnancy by both surgical and medical procedures.

2.1 Spontaneous abortion

The majority of embryos in archived collections, where the origin is recorded, were obtained after spontaneous abortion, the Kyoto collection (see below) being an obvious exception. A further significant number were surgically removed as treatment for ectopic implantation, mainly from the Fallopian tube, with smaller numbers obtained by therapeutic abortion, hysterectomy, or at post mortem. Spontaneous abortions are very common and hence provide a potentially accessible source of material with limited additional ethical implications, as the origin is a 'natural' process that has been widely investigated for many years. However, there is an underlying karyotype abnormality in up to 50% of cases (Hassold *et al.*, 1980; Warburton *et al.*, 1991). In addition a significant percentage of euploid aborted embryos have congenital abnormalities (Kalousek *et al.*, 1993), either overt or yet to become morphologically apparent, such as facial clefts that will only be evident at the end of the embryonic period (see later). These factors clearly limit the appropriateness of spontaneously aborted embryos as a source for study of normal human development.

2.2 Termination of pregnancy

Embryonic or fetal material can be easily collected at the time of therapeutic termination of pregnancy. Although frequently disrupted, embryos collected in this way are likely to be far more representative of a normal population than material obtained after spontaneous abortion. The incidence of cytogenetic anomalies in these essentially low-risk pregnancies is 4.5–6.3% (Yamamoto *et al.*, 1982; Zhou *et al.*, 1989). Combining the above data with incidences of various aneuploid and mosaic states in spontaneous abortuses and at the time of chorionic villous sampling (CVS), it has been possible to trace the natural history of these conditions (Wolstenholme, 1996). Cytogenetic work has also shed light on the various cell lineages emerging from the pre-embryonic period, their potential for differing genotype and outcomes with clinically important consequences, such as confined placental mosaicisms or uniparental disomy (Wolstenholme, 1996).

Embryonic and fetal tissue from terminated pregnancies have also been used to construct cDNA libraries. A variety of human fetal tissue cDNA libraries have been constructed in the past and are important contributors to the expressed sequence tag (EST) projects that underpin the efforts of the Human Genome Project to develop a catalogue of all human genes (Boguski and Schuler, 1995; Schuler *et al.*, 1996). They are widely available through commercial organizations such as Strategene, etc. Equivalent libraries from postimplantation human embryos are much rarer and are mostly whole-embryo cDNA libraries. One such has recently been reported by Jay *et al.* (1997) and the UK Medical Research Council is currently funding the construction of several others by the Newcastle Human Embryo Group in collaboration with the University of Oxford.

Surgical termination of pregnancy. Most of the above studies used embryonic material obtained by surgical termination of pregnancy, which is safe and widely employed in many countries worldwide. In England and Wales nearly 150 000 surgical terminations are performed each year (Office for National Statistics, 1996), of which 91% are undertaken at less than 13 weeks of gestation (11 weeks of development) by suction evacuation. Typically the anatomy of the embryo or fetus is severely disrupted during the procedure of suction termination, and although some embryonic parts may be identified for study, Carnegie Stage allocation will usually be by an unsatisfactory indirect method (Brailsford and Kaufman, 1982). It should be noted that the Kyoto University collection was largely obtained by a different method called dilatation and curettage. In Japan and elsewhere this technique has now been replaced by suction evacuation, which is quicker and has a lower frequency of complications (Beric and Kupresanin, 1971).

Medical termination of pregnancy. First trimester medical termination of pregnancy has been available in the UK since 1991, with the introduction of the progesterone antagonist mifepristone. Although not as popular as in continental Europe, it has provided a safe and effective alternative to the surgical procedure for women at or before 9 weeks of gestation, the limit imposed by the drug licence (UK Multicentre Trial, 1990). As most patients will abort the products of conception in the controlled environment of a hospital or clinic ward (compared with the obviously sporadic and unexpected nature of spontaneous abortion which usually occurs at home), this provides a relatively predictable opportunity for collection of early pregnancy material.

2.3 Study of the embryo in utero

The embryo *in vivo* has itself become the subject of research as technological advances allow the application of techniques previously developed for use in the fetal period. High frequency ultrasound using a transvaginal probe has allowed detailed longitudinal study of anatomical development, and measurement of parameters such as volume of the rhombencephalic cavity (Blaas *et al.*, 1995). In turn, these so-called sonoembryological studies have allowed the diagnosis of Meckel–Gruber, Joubert and Walker–Warburg syndromes on the basis of structural abnormalities of the hindbrain from as early as the end of the embryonic period (van Zalen-Sprock *et al.*, 1996).

3. Archived human embryo collections

To date, most work undertaken on human embryos has been at the level of anatomical observation, involving study of embryos in archived collections. The embryonic material in such collections has been used in some cases for foundation human embryology textbooks, as in the case of the Carnegie Collection in Washington, (O'Rahilly and Muller, 1987), the Boyd collection in Cambridge (Hamilton and Mossman, 1972), and the Blechschmidt collection currently housed at Louisiana State University, New Orleans but expected to be returned to Germany in the near future (Blechschmidt, 1963, 1973). The Carnegie collection, in particular, has been a useful reference source, being the foundation of the widely used Carnegie staging system for subdividing the first eight weeks of human development into 23 stages, up until the end of the embryonic period (O'Rahilly and Muller, 1987; see also Chapter 3). This collection, which was initiated in 1887 by Franklin P. Mall has been compiled from a variety of sources, and has accession records for 10 299 specimens, of which approximately 7000 are represented by histological sections or blocks. The collection includes a total of 695 embryos which have been fully sectioned, but which are of varying quality. A subset of 660 of these embryos were developed by George Streeter and Ronan O'Rahilly and used as the reference source for devising the 23 Carnegie stages (O'Rahilly and Muller, 1987). The sections in this and other archived embryo collections have been mostly stained for histological assessment, and they are not suitable for subsequent gene expression studies.

Another particularly influential collection was initiated in the 1960s by Nishimura and colleagues at Kyoto University, Japan, and comprises approximately 42 000 human embryos and fetuses, including over 7000 'well-preserved' specimens (Shiota, 1993). This series was largely obtained using the dilatation and curettage method, a procedure which causes much less embryonic disruption than standard suction evacuation (see above). It has provided unique information on the proportion of conceptuses with external major malformations, particularly holoprosencephaly, neural tube and cardiac defects, and facial clefts (Nishimura, 1975). Based on this collection Shiota (1993) calculated that approximately 10% of embryos at week 5 of development (Carnegie Stages 12–14) have, or will develop morphological abnormalities identifiable in the embryonic period. Of these, over 90% are destined to be spontaneously aborted before achieving viability. Unfortunately, karyotype information has not been recorded for this fascinating collection.

4. Ethical considerations

Ethical guidelines for use of embryonic (non-IVF) and fetal tissue for research in the UK have been established by the Polkinghorne Committee (Polkinghorne,

1989), set up by the Human Fertilisation and Embryology Authority. The main recommendations stress, first, the separation of granting the patient's request for termination from the request by the clinician for use of the embryonic or fetal tissue. Secondly, the clinician performing or procuring the termination should not be the researcher directly using the material obtained. Thirdly, confidentiality also needs to be rigorously maintained, and should not be put in jeopardy by separate records kept in the research institute pertaining to patients from whom material is collected. Local ethical committee approval is required for all such clinically related research, and this body should be satisfied that the above guidelines will be met.

The ethics of termination of pregnancy are beyond the scope of this chapter (see Chapter 4 for one perspective). Social expectation in the UK is that this option should be readily available to women. The gynaecologist's role is relatively passive and non-judgmental, with few requests for termination denied. There is a legal requirement for a second clinician (often the patient's general practitioner) to consider the patient's request and also give a signature of agreement (*Abortion Act, 1967*). The requirements of the *Abortion Act* and the Polkinghorne Report diminish the possibility of a researcher performing inappropriate termination procedures in order to obtain embryos.

5. The need for new human embryo collections

After a hundred years of ordered and meticulous study of descriptive embryology, the mechanisms of development and its control are now being slowly unravelled at a new, molecular level. Extrapolation of information on the genetic mechanisms controlling early development from animal models has been necessary because of the practical and ethical problems of human embryo research. Furthermore, it has been justifiable on the basis of observations that many developmental processes, and indeed developmental control genes, have been strongly conserved during evolution. However, there are now persuasive arguments to suggest that future developmental research should include studies in human embryos (Burn and Strachan, 1995; Strachan *et al.*, 1997; and see Chapter 2).

Advances in human structural genomics, together with the availability of new molecular tools such as tissue *in situ* hybridization and *in situ* reverse transcriptase-polymerase chain reaction (RT-PCR), will allow high resolution gene expression studies to be undertaken, which can provide detailed spatiotemporal expression data. Such profiles, and the observation of sequential, overlapping or abutting expression patterns, are likely to provide new insights into gene function. Early studies of gene expression in human and rodent embryos are revealing distinct species differences for some expression domains (El-Amraoui *et al.*, 1996; see also Strachan *et al.*, 1997). Future studies are likely to reveal more species differences, offering explanations as to why the phenotypes of certain animal disease models do not resemble their human counterparts. Likewise, as more human disease-related genes and plausible candidate genes are identified, detailed expression studies will have an important role, as illustrated by a recent report in which a candidate gene, *TBX5*, for Holt–Oram syndrome was identified (Li *et al.*, 1997).

Tissue *in situ* hybridization studies on human embryos revealed expression in the forelimb and heart, the organs characteristically affected in this disorder. Furthermore, high levels of expression were detected in the retina and anterior thoracic wall, potentially accounting for the reports of ocular and pectoralis major defects in a proportion of individuals with Holt–Oram syndrome.

Therefore, given the shortcomings of mammalian models (see Chapter 2), and the advances in means to investigate and explain the basis of clinical developmental disorders, there is an increasing need to systematically collect and store human embryos to supplement animal and *in vitro* studies.

6. The Newcastle Human Embryo Collection

6.1 Background and ethics

The Newcastle Human Embryo Group, which is managed by Tom Strachan, Susan Lindsay, David Wilson and Steve Robson, has been collecting human postimplantation embryos in order to obtain material for molecular genetic and immunohistochemistry studies of the molecular basis of early human development. The investigative techniques we have been using, particularly tissue *in situ* hybridization, depend upon intact anatomical architecture in order to provide precise spatial localization of molecular events (mRNA expression), and hence require morphologically intact embryos. We are compiling a human embryo collection, unique in that the embryos originate from apparently normal on-going pregnancies, are mostly karyotyped, are prospectively assigned to a Carnegie stage, and are handled and stored to preserve mRNA from degradation. In this section we describe our experiences using two methods for collecting embryos from patients undergoing termination of pregnancy. Intact embryos have been retrieved from the third week postconception to the end of the embryonic period (Carnegie Stages 8–22 so far). This encompasses the events of neural tube closure, and almost all the period of organogenesis, up to the end of the embryonic period (days 26–56 of development).

Burn and Strachan (1995) have recently reviewed the ethical considerations relating to the use of human embryos in developmental research, emphasizing the importance of the Polkinghorne report regarding the research use of human fetal and embryonic tissue in the UK (Polkinghorne, 1989). In keeping with the recommendations of this report, individuals involved in the decision to terminate or the termination procedure itself were not permitted to participate in research on the embryonic material collected in Newcastle. With these requirements in place our ethical committee have endorsed the retention of embryonic and fetal material for medical research, providing that such tissues are treated in a respectful fashion and that individual research projects are scientifically valid.

For the purposes of our embryo collection work all women undergoing routine medical or surgical termination of pregnancy are given an information sheet about the use of embryonic material for medical research. It is explained that agreement to participate in the research entails no added risks. Written consent is obtained from those likely to be 10 weeks gestation (8 developmental weeks) or less who agree to participate. The study was approved by the Joint Ethics

Committee of Newcastle and North Tyneside Health Authority and the University of Newcastle upon Tyne. In our experience about 80 % of women having a termination of pregnancy agree to allow embryonic material to be collected and retained for medical research.

6.2 Embryo collection methods

Embryonic material in the Newcastle Human Embryo Collection has been retrieved following both surgical and medical termination of pregnancy. We detail below the normal procedures that are followed to collect embryonic material.

Embryo collection following surgical termination of pregnancy. Cervical priming with prostaglandin E2 (3 mg) is performed for nulliparous patients thought to be 9 weeks gestation or more. All women agreeing to embryo capture have an abdominal ultrasound scan performed after induction of anaesthesia. No specific bladder preparation is undertaken. The presence of an embryo is confirmed, and the crown–rump length (CRL) measured.

The distal tip of a soft plastic 12 mm diameter suction curette (Rocket, London) is modified with a scalpel blade to give a smooth, angled oval opening. A 10 mm curette is suitable for embryos of 15 mm or less, although the tip may need to be cut to a slightly more pointed shape in order to penetrate the membranes of smaller sacs. The other end of the curette is attached to a 50 ml bladder syringe containing enough saline to fill the dead space of the curette.

The cervix is then dilated in the normal way for this procedure, and the plastic curette introduced into the uterine cavity under ultrasound guidance. Decisive early breach of the membranes prevents undue stripping of the chorion from the uterine wall which often predisposes to embryo disruption. Failure to enter the chorionic sac is usually discernible on ultrasound, but can be checked by the presence of resistance when aspiration with the syringe is attempted. Once inside the chorionic sac, the tip of the curette is placed beneath one pole of the embryo, and gentle negative pressure generated with the attached 50 ml syringe. In most cases this results in the embryo being drawn into the curette which is then removed from the uterus. The embryo can then be gently expelled into a sterile collecting vessel and fixed immediately. The termination of pregnancy is then completed in exactly the same way as usual by vacuum aspiration. The entire attempt at embryo capture typically takes less than 2 min.

Initial attempts at capture of late embryos (i.e. Carnegie Stage 23), or early fetuses with a CRL of 30 mm or more on ultrasound assessment, consistently resulted in significant disruption and so capture attempts are now limited to those with a CRL less than 30 mm. In a 24 month period between May 1995 and the time of writing (May 1997), surgical aspiration was attempted in 106 cases with a CRL measurement of less than 30 mm. Embryos were obtained from 64 (60%) attempts, of which 28 (26%) were intact. Although the intact embryos ranged from Carnegie Stage 12 to 22, a single Stage 12 embryo was the only intact specimen obtained prior to Stage 16 (*Figure 1*).

Embryo collection following medical termination of pregnancy. Medical termination of pregnancy in the first trimester is available up to 9 weeks gestation,

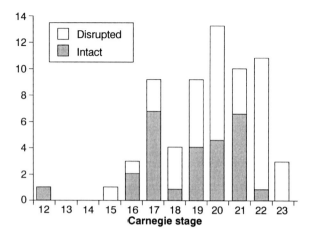

Figure 1. Representation of Carnegie stages in embryos recovered following surgical termination of pregnancy. Crown–rump length by ultrasonography <30 mm, n =64.

the limit determined by the drug licence for mifepristone. Women considering this option for termination have an ultrasound scan in the out-patient clinic, and are suitable if the CRL is 20 mm or less. Termination is procured by 200 mg of mifepristone taken orally as an out-patient, combined about 48 h later with 800 µg misoprostol self-administered vaginally. We initially used a dose of 600 µg mifepristone but achieved comparable results with the reduced dose. Most women abort whilst on the day ward between 2 and 6 h after administration of the prostaglandin. All blood and products of conception passed vaginally are collected for inspection as soon as possible afterwards. Embryos and apparently whole gestational sacs are separated out and placed immediately in 4% paraformaldehyde (PFA) or phosphate-buffered saline (PBS), for transport back to the laboratory for stereomicroscope (Zeiss) assessment and dissection where required.

Collection from medical terminations was attempted in 210 cases between December 1995 and May 1997 (18 months). Eighty six embryos (41%) were obtained, of which 51 (24%) were intact. The undamaged embryos spanned Carnegie Stages 8–21 (*Figure 2*).

6.3 Embryo processing, storage and assessment of morphology and karyotype

Recognized techniques are employed for reducing the effects of both endogenous and exogenous ribonuclease activity upon the collected specimens. The latter includes early separation from contaminating body fluids, embryo handling with surgical instruments and glassware that have been appropriately baked, and work surfaces swabbed with 0.1% diethylpyrocarbonate (DEPC) in ethanol. Endogenous ribonucleases are countered by commencement of fixation at the earliest opportunity, or liquid nitrogen storage where appropriate. Intact and mildly disrupted embryos are routinely fixed in 4% PFA, wax-embedded and subsequently stored at 4°C. Embryos that are more significantly disrupted are usually stored in liquid

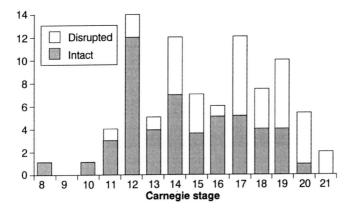

Figure 2. Representation of Carnegie stages in embryos recovered following medical termination of pregnancy. *n=86.*

nitrogen for subsequent isolation of RNA (and thence cDNA), having initially been placed in ice-cold PBS. The length of time between recorded passage of tissue until commencement of fixation or freezing is routinely recorded.

Assessment of morphology. Stereomicroscope assessment of Carnegie stage using external morphological features (see Chapter 3) is performed as soon as possible after collection. Photography and image capture records are made at the same time, and both performed rapidly for embryos chosen to be stored in liquid nitrogen. Embryos are defined as intact if there is no evidence of physical disruption on stereomicroscopic examination. A careful search for major malformations, particularly neural tube, craniofacial and limb defects, is routinely undertaken in all intact embryos. The following morphological abnormalities were confirmed in four of 79 (5%) intact embryos: microcephaly (Carnegie Stage 20), heart tube and outflow tract anomalies with branchial arch hypoplasia (Stage 12) and two neural tube defects (Stages 14 and 20). One further case of encephalocele may yet be confirmed by histology.

Karyotype analysis. We now routinely submit trophoblast tissue for cytogenetic analysis. Thus far, karyotypes have been obtained by cytogenetic assessment of trophoblast samples from 57 capture specimens where an intact or minimally disrupted embryo was noted to be present. Three (5.3%) abnormal results have been noted to date; one 45X, and two mosaics 46XX/47XX+2 (19 : 1 cells) and 46XX/47XX+16 (6 : 4 cells).

7. Discussion

7.1 Embryo collection

Both medical and surgical termination of pregnancy allow the successful collection of intact human embryos for research purposes. In the experience of the

Newcastle centre the vast majority of women agree to allow an attempt at embryo capture following a sensitive explanation of the potential importance of the research work being undertaken. This has allowed the acquisition of intact embryos covering a wide and critical window of human development including closure of the neural tube (Carnegie Stages 11–12), and the entirety of the period of organogenesis (with the exception of intact brain at Stage 23 so far). Indeed if the acquisition of currently single embryos at Carnegie Stages 8 and 10 proves to be repeatable, then even the earliest events of heart and central nervous system development will be accessible to investigation at a molecular level.

In our experience, embryo collection was numerically more successful using the surgical rather than medical method [64/106 (60%) versus 86/210 (41%) respectively]. This difference reflects the fact that some women abort after (and some before) the time spent in hospital, and material cannot always be saved for examination. However, the proportion of attempts yielding an intact embryo was not significantly different [28/106 (26%) versus 51/210 (24%) respectively]. In addition, collection of specimens from medical termination of pregnancy has given the greatest range of Carnegie stages (8–20). The upper range of developmental stages for embryonic material obtained using the medical termination of pregnancy is dictated by the drug licence for mifepristone, allowing use up to 9 weeks of gestation (49 developmental days), which corresponds approximately with Carnegie Stage 19. The lower limit is determined largely by logistical considerations – that is, the time between confirming a pregnancy after a missed menstrual period and attending for a termination. It is surprising, therefore, that embryos as early as Carnegie Stage 8 (about 16 days postovulation) and Stage 10 (21 days) have been obtained. Both were found in an appropriately grown gestation sac (proportional to the embryo) and had a normal karyotype, suggesting development was proceeding normally. Uncertainty about menstrual dates seems to have led to the unusually early attendance in both cases.

The surgical aspiration technique used is similar to that reported by Soothill and Rodeck (1994). They described ultrasound-guided aspiration of a single early fetus with trisomy 18. The technique requires experience of ultrasound-guided procedures, but in experienced hands can be completed in 1–2 min. The proportion of intact embryos collected does appear to improve with increasing experience. The opportunity for capture of early embryos seems more limited, because although the full range of intact specimens to date spans Carnegie Stages 12–22, only one embryo is actually less than Stage 16. Poor abdominal ultrasound resolution is the major factor limiting the range of embryos that can be captured by this technique. Although transvaginal imaging improves resolution of early embryos, the presence of the ultrasound probe makes the operative procedure extremely difficult. We have never successfully collected an intact embryo using unguided aspiration at the time of suction evacuation. In contrast, using the technique of dilatation and curettage without ultrasound, 18% of the Kyoto collection (6212 embryos and 1142 fetuses) were stated to be 'well preserved'. We found the upper size limit of collection to be determined by the 12 mm diameter aspiration curette. The largest embryo we have captured intact had a CRL of 27 mm (Carnegie Stage 22), and although most others collected of this size have relatively minor head disruption, this becomes consistently more marked at a CRL of 30 mm or more. The use of a larger aspiration curette may allow capture of intact embryos above Carnegie Stage 22, but passage would require additional cervical dilatation. The brief procedure outlined has not been found to cause any additional morbidity and indeed the use of ultrasound which is not routinely available for surgical termination, has been shown to reduce the risk of retained products of conception and other complications (Fakih *et al.*, 1986).

A potential concern about embryos collected from medical termination is the uncertainty as to when embryonic demise actually occurs, and hence whether autolytic damage occurs prior to fixation. Autolysis may then reduce RNA retention, producing suboptimal expression studies. A practical delay between expulsion and placement in fixative is also inevitable because of the uncertain nature of when expulsion will occur. Most intact embryos collected, however, were retained within an intact chorionic sac, presumably providing protection from exogenous sources of ribonuclease (maternal blood) and other degrading influences. Reassuringly, our experience (Li *et al.*, 1997; and see Chapter 11), and that of other groups that have undertaken gene expression studies on such embryos (Gerard *et al.*, 1995; and see Chapters 10 and 12) is that mRNA is amply retained, certainly in embryos that have remained intact through the process of expulsion. Not only have tissue *in situ* hybridization experiments been successful on embryos collected following medical termination of pregnancy, but direct comparison with embryos from surgical collection shows no apparent differences in target mRNA detection of presumed low copy genes such as members of the *WNT* family. The surgically captured embryos used for comparison are known to be fixed within a few minutes of the termination being performed.

7.2 Morphological abnormalities in collected embryos

External morphological anomalies were present in 5% of the undisrupted embryos we collected which have a median Carnegie Stage of 16 — the sixth week of development. In the large Kyoto University collection, Shiota (1993) calculated a risk of malformation of 9.7% for embryos in the sixth week postfertilization, including those abnormalities, such as cleft lip or palate, that were not evident at this time. Holoprosencephaly was one of the commonest anomalies reported, with a prevalence of 0.7% in specimens of Carnegie Stages 14–23 (Nishimura, 1975). We have not yet found an example of holoprosencephaly, although many of our embryos have not yet been sectioned.

It is worthy of note that with continuing advances in prenatal diagnosis it is likely that structural anomalies will increasingly be detected by early (transvaginal) ultrasound. Pathological examination of surgically aspirated conceptuses is likely to become an important means of confirming the prenatal diagnosis, particularly in those conditions where post-termination chromosomal or DNA studies cannot be used (van Zalen-Sprock *et al.* (1996).

It is now our policy to perform placental karyotyping on all collected embryos. The incidence of karyotypic anomalies in our small series (5.3%) makes this a necessity given our aim of studying gene expression, and establishing apparently normal parameters. This figure is in keeping with the larger studies in low-risk pregnancies cited earlier, and also the figure calculated by Hook's 'life tables', based on epidemiological data on spontaneous abortion (Hook, 1981).

6

Mouse and human embryonic development: a comparative overview

Matthew H. Kaufman

1. Background

The principal aim of this chapter is to provide an overview of murine and human embryonic development. Accordingly, it was decided that this could best be achieved by setting out, as briefly as possible, the major temporal changes that occur during human embryonic and fetal development, while in parallel emphasizing the similarities and differences that exist between these two species. Not all readers will be familiar with the terminology that has been established over the years and is essential for describing the detailed anatomy of the mammalian, but particularly the human, conceptus, therefore all of the anatomical terms used are highlighted in **bold** case when they first appear in the text. It is hoped that this approach will be of particular value to molecular and developmental biologists many of whom may not be familiar with the standard terminology used.

A brief overview is provided of the major developmental and configurational changes that occur in the human embryo during the period of embryogenesis. This is followed by an account of the principal events that occur during organogenesis up to approximately the end of week 10 of gestation. Limitations of space necessitate that only those events that involve the major organ systems, namely the nervous system and its closely allied structures, such as the eye and the ear, the heart and cardiovascular system, the limbs, the gastrointestinal and urogenital tracts, and the derivatives of the branchial arch system are considered, as these are the principal organ systems that are developing over this period. The external genitalia, for example, are only briefly mentioned, as their development is largely initiated towards the end of the period in question. Similarly, no attempt is made to discuss the principally genetic factors which influence early mammalian development, as this is also beyond the scope of this chapter.

The information provided in this chapter has been largely gleaned from an analysis of material published in standard textbooks of human and mouse embryology. The former, or comparable texts, are readily available for consultation in

Molecular Genetics of Early Human Development, edited by T. Strachan, S. Lindsay and D.I. Wilson.
© 1997 BIOS Scientific Publishers Ltd, Oxford.

most biomedical libraries should a more detailed analysis than that provided here be necessary. Those books that were found to be particularly helpful to the author were: Gasser (1975), Hamilton and Mossman (1972), Larsen (1993), Moore and Persaud (1993) and O'Rahilly and Müller (1987). With regard to the embryology of the mouse, the following are the standard sources of information on this species: Kaufman (1992), Otis and Brent (1954), Rugh (1968, reprinted 1990), Snell and Stevens (1966) and Theiler (1972, reprinted with minor changes, 1989). There is close similarity between the early development of the rat and mouse, and therefore a number of invaluable reference texts which relate specifically to the development of the rat should also be noted: Hebel and Stromberg (1986), Keibel (1937), and Witschi (1962), while Hamburger and Hamilton (1951) remains the standard reference work on the development of the chick embryo. For additional information on the staging of various vertebrate species, see also Butler and Juurlink (1987).

Standard developmental biology texts, by contrast, have always tended to concentrate on the underlying principles involved (see, for example, Balinsky, 1981; Bard, 1990; Gilbert, 1994; Thompson, 1917; Waddington, 1956; Weiss, 1939), and traditionally have devoted little or no space to descriptive accounts of the early stages of differentiation that occurs during the pre- and early postimplantation stages of mammalian embryonic development, or to the detailed events associated with the differentiation of their various organ systems. As a result, even experienced research workers within the fields of molecular and developmental biology, who may be required to have a detailed knowledge of the anatomy of the early mammalian embryo, are often particularly poorly qualified to interpret their *in situ* histological sections, especially if these relate to developmental stages and the detailed histological morphology of organ systems which lie outside their immediate area of interest.

Information from recent clinical studies using hormone immunoassay techniques to determine the time of the onset of human pregnancy, and advances in medical imaging and ultrasound techniques, now allow a more accurate means of determining the gestational age of a conceptus than was previously possible. These various approaches thus supplement earlier means of calculating the gestational age of the early and mid-gestation human conceptus as well as more advanced pregnancies from a knowledge of the date of the last menstrual period (if available) and from an analysis of the uterine volume made from a relatively inaccurate determination of the level of the uterine fundus.

1.1 Developmental and other subdivisions of human gestation

There are no generally accepted 'standard' ways of subdividing the 38–40 weeks of human pregnancy, beyond the pre-embryonic, embryonic and fetal periods. While this terminology has some advantages, it is equally valid, and probably more useful, to subdivide it into phases which more closely describe the *developmental* events that occur during this period. The first or 'pre-embryonic' phase, covers the period from fertilization (syn: conception) up to about day 14 *post conception* (p.c.). This almost exactly coincides with the process of gastrulation with the establishment of the intra-embryonic mesoderm and shortly afterwards the appearance of the primitive streak, being the first indication of the embryonic axis.

The coining of the term 'pre-embryo' is of relatively recent origin, and has legal rather than embryological connotations, being applied to the initial phase of human development during which, in the United Kingdom at least, a limited (but as yet not clearly defined) range of experimental manipulative studies may be carried out; for example, human sperm may be used to 'fertilize' zona-free hamster eggs in order to determine the genetic constitution of individual sperm, a technique now commonly used in infertility studies (Yanagimachi *et al.*, 1976), but the development of hybrid embryos, formed from the amalgamation of human and non-human gametes, is not acceptable. Similarly, while the establishment of human embryonal stem cell lines is probably acceptable, it has yet to be established whether they could be used to 'rescue' genetically deficient embryos.

It is during the first half of the 'pre-embryonic' phase, when the embryo possesses relatively few cells, but even up to the expanded blastocyst stage when 60–120 or more cells may be present, that the conceptus may be manipulated *in vitro* and, if required, individual cells or groups of cells may be isolated and analysed, either biochemically or genetically, while, under certain circumstances, allowing the rest of the embryo to continue developing (Adinolfi and Polani, 1989; Gardner and Edwards, 1968; Singer *et al.*, 1990). In the case of the human embryo, these techniques may only be carried out under strictly controlled guidelines (in the United Kingdom and in most other countries), and are invariably carried out in association with *in vitro* fertilization. The same constraints clearly do not apply to the embryos of other mammalian species, although in many but by no means all countries, experimental animal studies are carefully monitored. In the United Kingdom, for example, the latter are regulated by the *Animals (Scientific Procedures) Act 1986*, and are monitored by the Home Office Inspectorate. The numerous experimental studies that have been carried out over the years, with preimplantation mouse embryos in particular, have shed an enormous amount of light on the principally genetic factors that control early mammalian embryonic development (Bard and Kaufman, 1994).

During the second part of this first phase of development, once the embryo has implanted, the human embryo is completely inaccessible to experimental manipulation. In the mouse, by contrast, while implanting embryos are almost impossible to isolate without damaging them, developmentally slightly more advanced stages (e.g. egg cylinder or primitive streak stage embryos) are relatively easily isolated (Hogan *et al.*, 1994) and may be maintained in tissue culture for up to about 48 h, by which time many embryos would be expected to have achieved the limb-bud stage of development (Copp and Cockroft, 1990).

The second *developmental* phase continues until just beyond the beginning of the so-called 'fetal' period, and covers both the processes of embryogenesis and the majority of organogenesis. While by current usage (see O'Rahilly and Müller, 1987) the human *embryo* becomes transformed into the *fetus* at about the end of week 8 p.c. (this time was selected because it coincides with the first appearance of centres of ossification in the humerus), a more convenient *developmental* endpoint for this second phase might be a week or two later, that is, towards the end of week 10 p.c. By this time, the major events associated with organogenesis are close to completion, although in the majority of the organ systems the final stages of differentiation during which their detailed structural (i.e. histological) morphology is laid down may not be achieved until much later in gestation, and in some systems may not even occur until some years after birth (see below).

The third *developmental* phase embraces the period between the beginning of week 11 and full term, although the brain and the rest of the central nervous system and the respiratory system, for example, continue to differentiate for some considerable time after birth. While most of the major organ systems have yet fully to complete their differentiation by the end of the second phase, the third *developmental* phase essentially represents a period of consolidation and growth. It is the extended duration of this period in the more advanced mammals, and in the primates in particular, that largely distinguishes them from the other higher vertebrates. This difference is particularly evident when a comparison is made between the degree of development achieved in the mouse fetus at the time of delivery (on or about day 20 p.c.) with that achieved by the human fetus at term. Even by several weeks after birth, the mouse has yet to achieve the same degree of development *vis-à-vis* the differentiation of its various organ systems as the late-gestation human fetus. As an example to emphasize this point, in the mouse the eyelids close on or about 16 days p.c. and reopen approximately 12 days after birth (Findlater *et al.*, 1993), whereas in the human embryo the eyelids close at about 60 days p.c. (Pearson, 1980) while reopening occurs during the seventh month of intra-uterine life (Hamilton and Mossman, 1972), the underlying mechanism involved in each case being fairly similar.

The term 'pre-viable fetus', or an equivalent expression, is also commonly encountered, particularly in the earlier clinical and non-specialist literature, but over the years, principally because of recent advances in the care and management of early pre-term fetuses, this category of infants has become increasingly difficult to define. Formerly the legal dividing line was drawn at 28 weeks of gestation, so that a fetus that died before this stage of pregnancy was technically termed an *abortion*, while another whose gestational age at the time of delivery was between 28 weeks and full-term, but who nevertheless failed to survive, was technically termed a *stillborn*. This distinction is now more blurred because for a considerable number of years it has been possible to keep alive fetuses of as early as 24–26 weeks of gestation, although these infants, if they survive, have a high incidence of moderate to severe neurological damage as a direct consequence of their *prematurity* at the time of their delivery. The latter must be contrasted with the situation where survival is restricted, often only to mid-gestation, because of the presence of life-threatening congenital abnormalities (Kalousek *et al.*, 1990). Infants with life-threatening congenital malformations may, however, survive to term, but be stillborn, while yet others may die during the early postnatal period (the so-called neonatal deaths) (Bergsma, 1979; Winter *et al.*, 1988).

1.2 Availability of embryonic material suitable for analysis

In order to gain insight into the timing of the various developmental landmarks which are achieved by the human conceptus during these developmental phases, a variety of different approaches needs to be employed. While the preimplantation component of the first phase is accessible exclusively to those involved in *in vitro* fertilization and related procedures (Feichtinger and Kemeter, 1987; Seppälä and Edwards, 1985), the normal events that occur at and shortly after implantation are for most purposes completely inaccessible except to those who wish to examine isolated and appropriately 'fixed' and subsequently serially sectioned histological material (Gasser, 1975; Hertig *et al.*, 1956; O'Rahilly and Müller, 1987).

Relatively little serially sectioned human material is available which covers the peri- and early postimplantation period. The material available for study is also, for the most part, of relatively poor quality (mainly because of fixation problems, as it is hard to obtain fresh material) and this has inevitably led to difficulties in the interpretation of the detailed events that occur at this time. Equally, because of significant species differences in the peri-implantation events, extrapolation from observations made from the detailed analysis of the mouse to those events that occur during the comparable period in the human are of little value in relation to this exercise. This problem of species differences also applies to the interpretation of the detailed events that occur with regard to the differentiation of the extra-embryonic tissues during the very early postimplantation period when the trophectoderm of the conceptus interacts with the endometrial lining of the uterus. Even in the mouse, significant differences are observed even between closely allied rodent species (Hamilton and Mossman, 1972).

As indicated above, during the first trimester of pregnancy, human embryos *in utero* are largely inaccessible to the obstetrician as they progress through the second *developmental* phase as defined previously, although from about the fifth week of pregnancy modern ultrasonic visualization techniques may be of value in allowing the number of gestational sacs present to be accurately determined (Sabbagha, 1987). In some cases it may even be possible to establish whether the gestational sac is empty or contains a conceptus. From about 55 days p.c. until about 10–11 weeks p.c. invasive techniques may be employed to obtain small samples of extra-embryonic tissue suitable for DNA or cytogenetic analysis via the technique of transcervical or transabdominal chorionic villus sampling (CVS) or, somewhat later, usually between about 12 and 16 weeks of gestation, by amniocentesis. The latter technique, in addition to providing cellular material suitable for genetic analysis, also provides a sample of amniotic fluid which may itself be analysed biochemically. Early CVS is now only rarely undertaken because of the now-established risk of inducing craniofacial and/or limb abnormalities, termed the oromandibulofacial limb hypogenesis (OMFL) syndrome, in otherwise genetically normal infants as a direct consequence of this procedure, although the underlying mechanism(s) involved in inducing abnormalities of these systems has yet to be fully determined (Firth *et al.*, 1991; Kaufman, 1994). The spontaneous incidence of this condition is said to be in the region of 1 : 175 000 livebirths (Froster *et al.*, 1989), whereas after early CVS it appears to be in the region of between 0.09% (Schloo *et al.*, 1992) and 1.7% (Firth *et al.*, 1991).

It is, nevertheless, of critical importance that the embryologist and developmental biologist possess a reasonably detailed knowledge of the events that occur during the pre- and early postimplantation period. While the fate of the derivatives of the inner cell mass in the mouse and the human are nominally similar, the mechanism whereby the same end result is achieved in these two species at, for example, the stage when their embryos possess about 15–20 pairs of somites is quite different. This is principally because of the configurational differences that rapidly develop between mouse and human embryos and the quite different constraints imposed by their extra-embryonic tissues during the early postimplantation period. Attention is drawn to some of the principal species differences that exist between mouse and human conceptuses at this time in Section 3.

In the mouse, the degree of differentiation achieved on or about day 14.5 p.c. is approximately equivalent to that achieved by the human conceptus at the end of the so-called embryonic period. While a certain amount of information is available from the analysis of serially sectioned mouse embryos covering the period between days 14.5 and 18.5 p.c. (Kaufman, 1992; Theiler, 1972), much still needs to be learned of the detailed events that occur during this critical period. In the human, somewhat surprisingly, and in contrast to the situation in the mouse, no comparable complete reference collection of serially sectioned material is presently available. Indeed, it may well be unreasonable to expect that this deficiency in the availability of a human reference collection covering the complete *fetal* period is likely to be remedied in the forseeable future, principally because of the technical and ethical problems involved in obtaining suitable material for analysis.

The most complete reference collection of human *embryonic* material, the Carnegie collection, is presently located in the Armed Forces Institute, Washington, although smaller and consequently less representative but nevertheless well known collections of similar material are available elsewhere in the United States and in a number of European and other centres. The Carnegie collection contains in the region of 10 000 specimens (O'Rahilly and Müller, 1987), the majority of which are believed to be morphologically normal. Few, however, if any, were analysed cytogenetically at the time of their isolation, so that no information exists as to their chromosome complement. Since all of these specimens will have been obtained from either spontaneous or induced terminations of pregnancy, it is likely that a substantial proportion of the early embryos, in particular, will inevitably have an abnormal karyotype (Warburton et al., 1991).

It has certainly long been recognized that about 50–60% of all human spontaneous abortions have an abnormal karyotype, and that of these about 70% are aneuploid (these are mostly autosomal trisomics, while a significantly smaller proportion are X monosomics with an XO sex chromosome complement); a further 20–25% are polyploid, with a triploid : tetraploid ratio of about 3 : 1 (Boué and Boué, 1973, Therkelsen et al., 1973; Warburton et al., 1991). How this abnormal genotype is likely to have influenced their phenotype has yet to be determined, though embryos with an abnormal genotype invariably display a range of, sometimes extremely subtle, abnormal phenotypic features (Kalousek et al., 1990; Warburton et al., 1991).

1.3 The critical importance of the timing of the developmental events during early pregnancy

This chapter will emphasize the importance of understanding, albeit in fairly general terms, the *timing* of the developmental events that occur between implantation and about 10 weeks of gestation. This end-point approximately corresponds to the time when organogenesis is largely completed (see above). During this period critical events associated with organogenesis occur, and these are also, by their nature, particularly sensitive to a wide range of teratogenetic stimuli. While constraints on space do not allow pursuit of this particular line of discussion here, it is nevertheless of the greatest of importance in clinical practice to be aware of the detailed timing of the different stages of organogenesis, as this provides insight into the type and degree of severity of congenital abnormalities that may be expected to be induced should exposure to such stimuli occur during this period of development.

In the text that follows, the principal developmental landmarks that are observed in relation to the major organ systems are presented in an abbreviated form. Apart from the events that occur during weeks 2 and 3 which are considered in the context of the whole embryo, those that occur between weeks 4 and 10 have been treated separately for each of the major organ systems, as this allows their development to be relatively easily monitored. Such an approach also allows a rapid appraisal of the overall development of the embryo/fetus, viewed as a composite unit, to be made on an approximately week-by-week basis throughout this phase of gestation. Furthermore, it tends to emphasize the important periods when organ growth and differentiation are occurring, and consequently highlights the duration of maximum sensitivity of each of the organ systems to the effect of exposure to teratogenic insults, whether they be drugs, viral infections or ionizing radiation (*Figure 1*). It has to be emphasized, however, that the concept of a single sensitive period for each organ system is a considerable oversimplification. The reality of the situation is that the latter is the summation of numerous overlapping sensitive periods, and that exposure to a teratogenic stimulus during the early part of such a period would usually be expected to have a more detrimental effect on the differentiation of a particular system than if exposure to the same agent, in the same dose, had occurred towards the end of that sensitive period (Warkany, 1971; Wilson and Fraser, 1977).

The baseline information in this overview concentrates on the events that occur during human embryonic and early fetal development. This is principally because the prolonged gestation period of the human compared to the mouse, as well as the enormous literature on the descriptive events that occur during early human development, allows almost a day-by-day record of events to be presented here. This contrasts to the situation in the mouse where comparable events associated with organogenesis take place during a much more abbreviated period of time, and the relevant literature, although expanding rapidly, is also at the present time much less complete.

An overview is provided in *Table 1* which allows the principal features seen in mouse, rat, and human embryos during the early postimplantation period to be compared; in the mouse, this spans the period between implantation (day 4.5 p.c.) and the time when the lens vesicle completely separates from the surface ectoderm (day 11.5 p.c.); in the human embryo, the equivalent stages are observed on days 5–6 and 37 p.c., respectively. The major changes that take place in human and mouse embryos during the second part of the embryonic period are presented in *Table 2*. Frequent reference to these tables showing equivalent stages

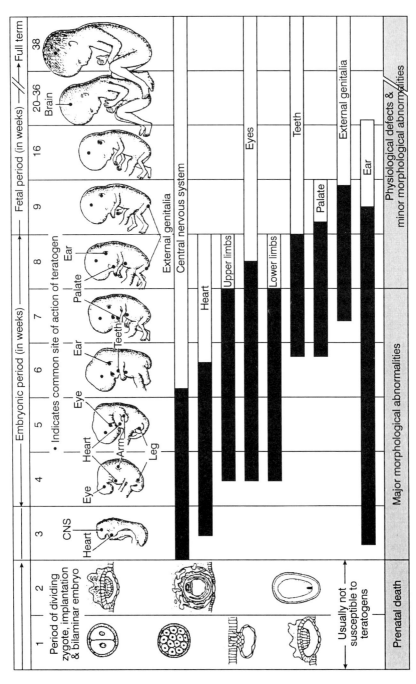

Figure 1. Schematic illustration of the critical periods in human development. Redrawn from Moore and Persaud (1993), with permission of W.B. Saunders Company.

Table 1. Timing of appearance of principal features in mouse and human embryos during the early postimplantation period

Principal features in rodents	Mouse				Human			
	Theiler's Stage	Age (days)	Size (mm) (unfixed)	Pairs of somites	Carnegie Stage	Age (days)	Size (mm) (fixed)	Pairs of somites
Attaching blastocyst	6	4.5			4	5–6	0.1–0.2	
Early egg cylinder. Ectoplacental cone appears. Embryo implanted, although previllous (human)	7	5			5	7–12	0.1–0.2	
Differentiation of egg cylinder. Proamniotic cavity appears	8	6						
Advanced endometrial reaction. Formation of mesoderm. Primitive villi (human, stage 6), branching villi (human, stage 7)	9	6.5			6, 7	13–16	0.2–0.4	
Amnion formation, primitive groove, allantois	10	7			8	18	1.0–1.5	
Neural plate, presomite stage	11	7.5						
First somites. Neural folds begin to close in occipital/cervical region	12	8–8.5		1–7	9	20	1.5–2.5	1–3
'Turning'. Two pharyngeal bars, optic sulcus	13	8.5–9		8–14	10	22	2–3.5	4–12
Elevation of cephalic neural folds, formation of rostral neuropore. Optic vesicle formation	14	9–9.5		13–20	11	24	2.5–4.5	13–20

(Continued on following page)

of embryonic development will be essential in order to gain a proper appreciation of the substantial time-scale differences that exist between comparable developmental events in these two species.

2. Principal events during the 2nd to the end of the 10th week of human embryonic and early fetal development

Week 2. By *day 10* p.c., the **blastocyst** becomes completely embedded in the endometrium; the **inner cell mass** differentiates to form the **bilaminar embryonic disc**; the **epiblast layer** is associated with the **amniotic cavity** (which appears during *day 8*), while the **hypoblast layer** is associated with the **primary yolk sac cavity** (which appears during *day 9*). The **extra-embryonic mesoderm** appears, and becomes located between the outer aspect of the **amnion** and the **yolk sac**, and the inner aspect of the **trophoblast**.

By *day 14*, differentiation of the **primitive streak** occurs. This represents the first indication of the embryonic axis, and establishes the bilateral symmetry of subsequent stages of development.

Week 3. During *days 15* and *16*, the critical morphogenetic event termed **gastrulation**, converts the **bilaminar embryonic disc** (in which two germ layers are present, namely **ectoderm** and **endoderm**) into the **trilaminar embryonic disc**; this event is associated with the movement of cells of **epiblast** origin (the primitive ectodermal lineage) into the **primitive streak**, and thence separates the overlying epiblast layer from the subjacent hypoblast layer (the primitive endodermal lineage) to form an intermediate layer – the **embryonic mesoderm**; its immediate cellular derivatives are mesenchyme cells, some of which displace the hypoblast layer (located across the ventral midline) laterally, to form the **embryonic endoderm**; the derivatives of the latter are the epithelial lining of the gut tube and its associated glands, and the lower respiratory tract.

Epiblast cells on the surface of the embryonic disc form the **embryonic ectoderm**; its derivatives are surface ectoderm/epidermis, neural ectoderm (which gives rise to the central nervous system and sensory epithelium of the eye, see below), the otocyst (which forms the inner ear apparatus, see below), and the olfactory epithelium (see below). Embryonic ectoderm also gives rise to the **neural crest** (which differentiates to form the coverings of the peripheral nervous system, the autonomic nervous system, the dorsal root ganglia and various components of cranial ganglia, the pigment epithelium of the body, and the connective tissue and skeleton of the face, etc.).

By *day 17*, cellular proliferation at the cranial end of the primitive streak forms **Hensen's node** (also termed the **primitive pit, node** or **knot**); while the latter is only poorly differentiated in mammals, it is well developed in avian embryos. The mesodermal cells that migrate rostrally through the primitive pit give rise initially to the **prechordal plate**, and this subsequently forms into a midline tube termed the **notochordal process**. The latter fuses with the subjacent endoderm to form the **notochordal plate** by *day 20*, and by *day 22*, this separates from the endoderm as the **definitive notochord**.

On either side of the primitive streak, the mesoderm spreads laterally, and during *weeks 3* and *4* forms three initially poorly defined columns of tissue. These differentiate in a craniocaudal sequence into a column immediately lateral to the primitive streak, termed the **paraxial mesoderm**; slightly lateral to this, a second column, but of less well-defined tissue forms, which is termed the **intermediate plate mesoderm**, and lateral to this forms another relatively poorly defined column termed the **lateral plate mesoderm**.

By *day 18*, the paraxial mesoderm, in its most rostral part, shows the first evidence of becoming segmented into the **somitomeres**; the **extra-embryonic coelom** (or **coelomic cavity**) forms following the splitting of the **extra-embryonic mesoderm**, so that the outer layer of the extra-embryonic mesoderm lines the trophoblast, while the inner layer surrounds the outer part of the walls of the amniotic and yolk sac cavities; the two components of the extra-embryonic mesoderm are attached by the **connecting stalk** (the future **umbilical cord**).

The derivatives of the **intra-embryonic mesoderm** are widespread throughout the body, and differentiate to form all of the muscle, connective tissue, cartilage and bone, as well as the heart and vascular system.

By *day 19*, the **neural plate** appears, and its cranial part gives rise to the future **brain**, while its more caudal part gives rise to all but the most caudal part of the future **spinal cord**.

By *day 21*, the **somites** (which are derived from the segmentation of the paraxial mesoderm) first appear, again in a cranio-caudal sequence, appearing first in the future occipital/upper cervical region.

By *day 22*, the first evidence of apposition and fusion of the neural folds to form the neural tube (the process of **neurulation**) is seen in the cervical region of the future vertebral axis, followed shortly afterwards by a cranial and caudal extension of this process over the period of the next few days to form the entire neural axis.

By *day 24*, the **cranial (rostral) neuropore** closes; the early differentiation of the rostral part of the neural tube allows the formation of the **three primary brain vesicles** (the **prosencephalon**, **mesencephalon** and **rhombencephalon** — the primitive **forebrain, midbrain** and **hindbrain** regions, respectively).

By *day 26*, the **caudal neuropore** closes, but the most caudal part of the neural tube forms by a less well organized and separate process than neurulation, termed **secondary neurulation**, in which the **caudal eminence**, formed from mesodermal elements, eventually fuses (during *week 6*) with the most caudal part of the neural tube. The lumen which develops within the caudal eminence subsequently fuses with the lumen of the neural tube.

By about *day 28/29*, dorsal (alar) and ventral (basal) columns differentiate within the **mantle layer** of the neural tube, while by about *day 30*, **dorsal root ganglia** are present at most levels along the neural axis. **Ventral roots** are initially found in the cervical region, and are subsequently seen to innervate the cervical myotome derivatives by about *day 33*; the first evidence of **sympathetic trunks** are also seen at about this time

Week 4. This is a critical one, with early evidence of the differentiation of the primordia of all of the major organ systems, and their development is discussed in the following sections.

2.1 Differentiation of the paraxial mesoderm

Differentiation of the **somites** and their derivatives largely takes place over the period between *weeks 4* to *8–10*. Initially, the principal derivatives of the somites are the **sclerotome** and **dermomyotome**, but the latter soon differentiates to form **dermatome** and **myotome**. Sclerotome cells migrate and surround the neural tube and notochord, and will subsequently condense to form the skeletal elements of the **vertebrae**, although the means by which this is achieved are somewhat complex.

Each somite forms at the same segmental level as a spinal nerve, but the sclerotome masses soon divide into cranial and caudal parts; the cranial part of one sclerotome mass (which consists of loosely packed cells) recombines with the caudal part of its immediately superior sclerotome mass (which consists of densely packed cells), so that at the end of this process the definitive vertebral elements are formed, each of which is located at intersegmental levels. The lower part of the original cranial half of the sclerotome mass differentiates to form the **annulus fibrosus** (the future **intervertebral disc**), in the centre of which is located the remnant of the notochord, termed the **nucleus pulposus**.

As a consequence of the latter phase-shift of half a segment, the segmental spinal nerves emerge at intersegmental levels, at the same level as the intervertebral disc, while the paravertebral muscle masses now pass from one vertebral body to its adjacent cranially and caudally located vertebral bodies, and by their muscular action can facilitate a limited degree of movement of the vertebral column.

The myotomes give rise to the epaxial muscles, while the hypaxial muscles give rise to the body wall musculature of the trunk, as well as the musculature of the limbs; the **dermatome** derivatives contribute to the dermis, and form the segmental basis for the cutaneous pattern of innervation of the skin.

2.2 Differentiation of the neural tube to form the brain and spinal cord, and closely allied structures: the early development of the optic, auditory and olfactory systems

The initial stages of the differentiation of the neural axis commence during the latter part of *week 3*, and are more clearly seen during *week 4*, with the process of **neurulation**, which converts the **neural folds** into the **neural tube**: the latter differentiates to form the **brain** (cranially) and the **spinal cord** (caudally). Additionally, **neural crest** cells separate from the region of the apices of the neural folds, and subsequently migrate to various parts of the body (e.g. to form the dorsal root ganglia and parts of many of the cranial ganglia). They also give rise to elements of the autonomic nervous system, including the sympathetic and parasympathetic ganglia, as well as a wide range of other non-neural elements — see above).

The closure of the rostral neuropore on *day 24*, also allows the formation of the **primary optic vesicles**; these constitute lateral evaginations of the future diencephalic region of the forebrain, although the **optic primordia** (the **optic sulcus** and **placode**) are first evident on about *day 22*.

The **otic placodes** (which will eventually form the **membranous labyrinth of the inner ear**) are first evident during *week 3*, and differentiate from a thickening of the surface ectoderm on either side of the future hindbrain. By the end of *week*

3, the otic placodes indent to form the **otic pits,** and subsequently (during *week 4*) completely lose their connection with the surface ectoderm to form the **otocysts** (or **otic vesicles**). The **pinna** of the external ear forms from the amalgamation of a series of about six **auditory** (or **auricular**) **hillocks** which form during *week 5* from surface ectodermal tissue of the first and second pharyngeal arches, which are located on either side of the first pharyngeal groove. The definitive appearance of the **external ear** is not achieved, however, until the newborn period. The first pharyngeal groove subsequently gives rise to the **external acoustic meatus,** while the **tubo-tympanic recess,** a derivative of the first pharyngeal pouch, will subsequently give rise to the cavity of the middle ear. It is only during the *ninth month* that the tubo-tympanic recess expands to enclose the **middle ear ossicles** and forms the definitive **middle ear cavity.**

The **olfactory** (or **nasal**) **placodes** develop on the **frontonasal process** during *week 4.* The central region of each of the olfactory placodes indents to form the **nasal pits,** which will (during *week 6*) eventually form the **nasal cavities.** The latter are initially continuous with the oral cavity, but with the fusion of the **palatal shelves of the maxillae** (during *week 10*), these are separated into the **oropharynx,** below, and the **nasopharynx** above the **definitive** (or **secondary**) **palate.** The primitive nasopharynx is separated into the two **nasal cavities** by the downgrowth of the **nasal septum** and its fusion in the midline with the dorsal surface of the secondary palate which also occurs during *week 10.*

In the cephalic region, during *week 5*, the prosencephalon becomes divided rostrally into the two **telencephalic vesicles,** while the more caudal part of the primitive forebrain is termed the **diencephalon.** The cavities within these parts of the forebrain are, respectively the telencephalic (or **lateral**) **ventricles** and the **third ventricle.** The rhombencephalon (whose cavity is termed the **fourth ventricle**) divides into cranial and caudal parts, termed respectively the **metencephalon** and **myelencephalon.** The cavity of the mesencephalon diminishes in volume to become the **cerebral aqueduct.** By this stage, the three **primary brain vesicles** have differentiated to form the five **secondary brain vesicles.**

The following regions of the brain are formed from the differentiation of the five secondary brain vesicles:

telencephalic vesicles — **cerebral hemispheres**
diencephalon — **hypothalamus, thalamus** and **epithalamus**
mesencephalon — **midbrain**
metencephalon — **pons and cerebellum**
myelencephalon — **medulla oblongata.**

By about *day 37*, the sympathetic trunks are seen to extend caudally into the thoracic region. The contents of the limb buds are seen to be innervated by segmental spinal nerves; these initially innervate the contents of the upper limb buds, and, slightly later, are also seen to innervate the contents of the lower limb buds. By about *day 43*, the vagal parasympathetic fibres grow out from the **parasympathetic ganglia** to innervate the heart. During *week 8*, outgrowths of the autonomic nerves are seen to be present at all levels along the neural axis.

Development of the eye. As indicated above, the **optic sulci** and **placodes** are first seen on or about *day 22*, while the **primary optic vesicles** form on *day 24*, and coincide with the closure of the rostral neuropore. By about *day 31*, the surface ectoderm overlying the optic vesicle thickens to form the **lens placode**. During *day 32*, this indents to form the **lens pit**, and by *day 33*, it has completely separated from the surface to form the **lens vesicle**. The cells of its posterior wall elongate, so that by about *day 47* the cavity of the lens vesicle is completely obliterated. The cells of the posterior wall of the lens vesicle form the **primary lens fibres**. Coinciding with the indentation of the lens pit, the optic vesicle also indents to form the **optic cup**. Its inner layer of cells forms the **neural layer of the retina**, while the cells of the outer layer of the optic cup form the **pigment layer of the retina**; the narrow gap initially present between these two layers is termed the **intra-retinal space**.

During *week 6*, the axons from the ganglion cells of the neural retina pass through the **optic stalks** to reach the brain. By *week 8*, the lumen of the optic stalks has become obliterated, and these are now termed the **optic nerves**. When the axons from one side reach the **optic chiasm**, about half of the nerve fibres pass ipsilaterally, while the other half pass contralaterally to reach the corresponding **lateral geniculate body**.

The surface ectoderm overlying the optic apparatus differentiates to form the **outer layer of the cornea**, while the **inner layer of the cornea** is formed from **cephalic neural crest**. The latter also condenses around the pigment layer of the retina to form the **choroid** and **sclera**. By the end of the *third month*, the anterior part of the optic cup and the overlying choroidal mesenchyme expand forwards to form a ring over the lateral part of the lens. This differentiates into the **iris** of the eye. Towards the end of *week 7*, neural crest also passes between the lens and surface ectoderm. It subsequently splits into two layers to enclose the **anterior chamber of the eye**. The **pupillary muscles** in the iris are also derived from neural crest of choroidal origin. The choroid also differentiates to form the **ciliary body**.

Both the developing lens and neural retina receive their blood supply (by *day 33*) from the **hyaloid artery**, a branch of the **ophthalmic artery**, which is itself a branch of the **internal carotid artery**.

The **extrinsic ocular muscles** are believed to differentiate from paraxial mesoderm, while the connective tissue associated with them is derived from neural crest. Their innervation is from the oculomotor (III), the trochlear (IV) and abducent (VI) cranial nerves, and occurs during *week 5*.

The **eyelids** develop during *week 6*, become apposed and fused during *week 8*, and usually reopen during the *sixth and seventh months*.

2.3 Differentiation of the heart

The first evidence of **vasculogenesis** is seen during *day 19*, in the **cardiogenic region of the embryonic disc**. By about *day 20*, two **endocardial tubes** are formed which amalgamate together across the midline in the future **pericardial region** of the **intraembryonic coelom**. By *day 21*, a midline **primitive heart tube** has formed, with subdivisions along its length. The cranial end forms the **aortic sac** and this is continuous with the first and second branchial arch arteries, while the caudal end develops into a **common atrial chamber** and receives the two horns of the **sinus venosus**. The wall of the primitive heart tube initially consists of **myocardial tissue**, subjacent to which is a layer of **cardiac jelly** which is believed to be secreted by the myocardial cells. The cardiac jelly is therefore located between the endocardial and myocardial layers of the heart. The **epicardium** forms from mesothelial cells possibly derived from either the wall of the sinus venosus or from the **septum transversum**.

During *day 22*, the heart begins to beat, while folding (or 'looping') of the primitive heart tube begins on *day 23*, and this process is effectively completed by about *day 28*. At about this time, the **septum primum** is first evident as a downgrowth from the roof of the **common atrial chamber**, and gradually grows in the form of a crescent towards the **atrioventricular endocardial cushion tissue**. By *day 38*, it makes contact with the cushion tissue, by which time part of the wall of the septum breaks down to form the **ostium secundum**. The **septum secundum** forms to the right of the septum primum, and is a thicker and more muscular structure than the septum primum. By about *day 42–43*, **intra-atrial septation** is completed, and the channel that allows communication between the right and left atria is termed the **foramen ovale**. This channel allows oxygenated blood from the placenta (mixed with a small volume of deoxygenated blood) to flow from the right atrium into the left atrium for distribution principally to the head and neck and upper limbs.

Interventricular septation is initiated at the end of *week 4* (on or about *day 30*), with the formation of the **muscular part of the interventricular septum**, which initially separates the primitive right from the primitive left ventricle. The more advanced development of the interventricular septum is coordinated with the differentiation of the **truncoconal septae** that develop within the outflow tract of the heart, the upper parts of which form the **aorticopulmonary spiral septum** which separates the future aorta from the pulmonary trunk. During *week 5*, the **truncoconal swellings** grow downwards, and subsequently fuse, during *week 9*, with the **atrio-ventricular endocardial cushion tissue** and the muscular part of the interventricular septum, thus separating the **definitive right ventricle** from the **definitive left ventricle**.

2.4 Differentiation of the limbs

The limb buds develop from the somatopleuric component of the **lateral plate mesoderm** in the flank region of the embryo, are covered by surface ectoderm, and contain a mesodermal core. The mesoderm, which gives rise to the limb musculature, is somite-derived, while the skeletal elements, tendons, ligaments and vascular components of the limb are derived from mesoderm of lateral plate origin.

The first evidence of **forelimb bud** development is seen in the future lower cervical/upper thoracic region on about *day 24*, while the **hindlimb bud**, which develops in the lower lumbar/upper sacral region, is first evident during *day 28*. Up to about *weeks 8–10*, the hindlimb is delayed by up to about a week in its differentiation compared to that of the forelimb, although, as pregnancy proceeds, the developmental difference between them diminishes.

Apical ectodermal ridge (**AER**) differentiation occurs along the peripheral margin of the limb bud, and is seen shortly after the first appearance of the limb buds. Its primary function (or that of the region immediately subjacent to it, termed the **zone of polarizing activity**) is to induce the mesenchyme within the limb to differentiate. It does so in a proximal to distal sequence, possibly through the release of a morphogen (which may be retinoic acid) so that the distance away from the influence of the AER appears to be the critical factor in facilitating the differentiation of the limb mesenchyme.

The principal role of the AER is therefore to control pattern formation within the limb, and this function is confirmed by the results of numerous experimental studies in which defined segments of the AER from embryos of various developmental stages have been transplanted into the distal parts of limb buds of embryos at the same or different stages of development, in order to determine the influence of the transplanted grafts. While these experimental studies were almost exclusively carried out with avian embryos, it is believed that the findings are equally applicable to mammalian embryos.

By about *day 33*, the forelimb is seen to be divided into a **handplate**, forearm, arm and shoulder regions, and similar subdivisions are seen in the hindlimb several days later. By *day 37*, the carpal region is now recognizable, while by *day 38*, the handplate is seen to have early evidence of **digital rays** which are separated by **digital interzones**. By *day 44*, the individual digits are clearly seen, due to progressive **programmed cell death (apoptosis)** in the digital interzones. By about *day 52*, **tactile pads** are evident on the palmar surface of the tips of the forelimb digits, and on the plantar surface of the distal parts of the hindlimb digits. By *day 47*, a moderate but progressive degree of medial rotation of the upper limb is occurring, and is maximally achieved by about *day 56*. In the hindlimb, lateral rather than medial rotation occurs, and is also maximally achieved by about *day 56*, and it is at about this time that both sets of limbs are seen to be well developed.

Within the limbs, the musculature is derived from somitic mesoderm which migrates into the early limb bud. The primitive muscle masses are initially innervated by **pathfinder nerves** which, in the forelimb, will eventually form the elements of the **brachial plexus**. In the hindlimb, the innervating nerves will form the elements of the **lumbosacral plexus**. In the central core of the limb buds, mesenchyme condenses to form the cartilage models of the future skeletal elements, while it is within the latter that the primary centres of ossification form, the process being termed **endochondral ossification**. Limb mesenchyme condenses during the *fifth week*, and ossification centres appear during *weeks 7–12*, the first bone to ossify being the clavicle (during *week 7*).

2.5 The gut tube and its associated glands

The endodermal gut tube extends from the **buccopharyngeal membrane**, cranially, to the **cloacal membrane**, caudally. Between these two boundaries are respectively the regions of the embryonic **foregut**, **midgut** and the **hindgut**. While the foregut and hindgut are relatively narrow in profile, in the region of the primitive midgut, the neck of the **yolk sac** progressively narrows and eventually forms the **vitello-intestinal** (or **vitelline**) **duct**, which extends from the mid-point of the midgut, distally into the umbilical cord. In a small proportion (about 3%) of the population, a remnant of the latter may persist as a **Meckel's diverticulum**.

The principal subdivisions of the primitive gut tube each have a discrete blood supply, with branches of the **coeliac trunk** supplying the region of the primitive foregut. The latter region includes the derivatives of the pharynx: including the lungs and lower respiratory tract [to be dealt with in the section on the 'pharyngeal (branchial) arch system'], the oesophagus, the stomach and the proximal half of the duodenum. The **superior mesenteric artery** supplies the midgut, which extends from the middle of the second part of the duodenum, and therefore includes the distal part of the duodenum, the jejunum and ileum, and the ascending and proximal two-thirds of the transverse colon. Branches of the **inferior mesenteric artery** supply the region of the hindgut, which extends from the junction between the proximal two-thirds and the distal one-third of the transverse colon, and includes the descending and sigmoid colon, the rectum and upper two-thirds of the anal canal. The primitive hindgut also includes the **urogenital sinus** (see below).

Up to the end of *week 4*, the primitive gut is represented by a straight tube which is suspended by a dorsal mesentery along much of its length, while the region where the stomach forms is the only part of the primitive gut which retains a ventral mesentery. The **septum transversum**, which develops within the ventral mesentery of the stomach, in due course gives rise to the parenchyma of the liver. The latter is suspended from the anterior abdominal wall by the **falciform ligament** and from the ventral wall of the stomach by the **lesser omentum**. During the embryonic/fetal period, the **left umbilical vein** (which carries oxygenated blood from the placenta) runs in the lower border of the falciform ligament. After birth, the lumen of this venous channel becomes obliterated, and the fibrous band that replaces it is termed the **ligamentum teres**.

The diameter of the stomach begins to expand during *week 5*, and is maximally expanded by the end of *week 7*. During *weeks 7 and 8*, the stomach rotates through 90° clockwise when viewed from above. Since the dorsal wall expands to a greater degree than the ventral wall, the stomach develops a greater (dorsal) and a lesser (ventral) curvature. The dorsal mesentery expands to form the **greater omentum**.

Aggregates of mesodermal tissue form in the **dorsal mesogastrium** during *week 5* to form the **spleen**, and this becomes suspended between the greater curvature of the stomach and the posterior abdominal wall respectively by the **gastrosplenic** and **lienorenal ligaments**. Up to about *week 14* of gestation, the spleen serves as an important organ of haematopoiesis, second only in importance to the liver in this regard, but from about *week 15*, it gradually becomes invaded by lymphoid tissue, many of these cells being T-lymphocyte precursors, and the spleen then increasingly takes on an immunological role.

During *weeks 4–6*, various diverticulae bud from the region of the foregut–midgut junction. One of these, the **biliary bud**, elongates and repeatedly branches within the tissue of the septum transversum to form the **intra-** and **extra-hepatic components of the biliary system**. Two other diverticulae which sprout from this location are the **ventral** (smaller) and **dorsal** (larger) **pancreatic buds** (or rudiments). These enlarge and eventually amalgamate to form the **definitive pancreas**, so that the **main pancreatic duct** is formed from the union of the proximal part of the duct of the ventral pancreatic rudiment which joins (distally) to the duct of the body and tail of the dorsal pancreatic rudiment. The ventral pancreatic rudiment forms the majority of the head and uncinate process, while the dorsal pancreatic rudiment gives rise to the rest of the body and tail of the definitive pancreas. An **accessory pancreatic duct**, if present, is formed when the main duct of the dorsal pancreatic rudiment is retained instead of disappearing, as is normally the case. It is for this reason that both the common bile duct and the main pancreatic duct normally enter the middle of the second part of the duodenum, at the foregut–midgut junction, at a single site, the **greater duodenal papilla**.

During *week 7*, the midgut expands enormously in length and, because there is insufficient space within the peritoneal cavity for this to occur (the liver, in particular, and stomach occupy the upper half of the peritoneal cavity, while the **mesonephroi** expand forwards to occupy much of the rest of the available space), it herniates into the proximal part of the umbilical cord, into the so-called **physiological umbilical hernia**. Within the latter, the midgut loop initially rotates through 90° anticlockwise, when viewed from the front, and then, as it returns into the peritoneal cavity, during *weeks 10–12*, it rotates a further 180° anticlockwise. Once the midgut has fully returned into the peritoneal cavity, after rotating through a total of 270°, its various parts have by this time assumed their definitive locations. The ascending and descending regions of the colon become 'fixed' to the posterior abdominal wall, while the rest of the midgut is suspended by its dorsal mesentery.

The midgut is able to return into the peritoneal cavity because the potential space available within the peritoneal cavity has increased over the period between *weeks 10* and *12*, principally because the rate of increase in volume of the liver diminishes compared to that of the peritoneal cavity, and because the mesonephroi have largely regressed and been replaced by the considerably smaller **metanephroi** (or **definitive kidneys**).

The distal part of the hindgut gives rise to the **rectum** and upper two-thirds of the **anal canal** (dorsally), and the **urogenital sinus** (ventrally), and these are subsequently partitioned by the downgrowth of the **urorectal septum** during *weeks 6–7*. The primitive cloacal membrane accordingly becomes subdivided into the **anal membrane** and **urogenital membrane**, the two being separated by the **perineal body**. All of the derivatives of the cloaca are supplied by the pudendal nerve. The junctional zone between the primitive hindgut and the ectodermal invagination, the **anal pit** (or **proctodaeum**) is at the **pectinate line**. The anal membrane normally breaks down during *week 8*. While the site of the junction between the hindgut and the anal pit is well delineated, that between the **oral pit** (the **stomodaeum** or **stomatodaeum**) and the foregut, at the site of the former buccopharyngeal membrane, is not.

2.6 Development of the urogenital system

Both the urinary/renal and the internal genital duct systems develop from **intermediate plate mesoderm**, which is first recognized at about *day 19*, shortly after the establishment of the three germ layers. By *day 24*, segmentation occurs in the most rostral part of this column of tissue, and this gradually extends caudally. The most rostral elements constitute the **cervical nephrotomes**, and their features suggest that they are equivalent to the primitive kidneys (or **pronephroi**) seen in fishes. In man, they are non-functional, and regress during week 4, being replaced by the **mesonephroi**. These represent 'intermediate' forms between the 'primitive' kidneys (the pronephroi) and the 'definitive' kidneys (the metanephroi). They are extensive structures (the **Wolffian bodies**) which are located in the thoracic and lumbar regions, and are believed to function in man. They are drained by a pair of ducts (the **mesonephric ducts**) which open into the posterolateral part of the **urogenital sinus**.

By *week 5*, a small diverticulum (the **ureteric bud**) extends rostrally from near to the site where the mesonephric duct enters the urogenital sinus. The ureteric buds make contact with the most caudal part of the intermediate plate mesoderm-derived tissues, and induce them to develop into the **metanephroi** (or **definitive kidneys**). Repeated branching of the tip of the ureteric bud occurs within the metanephros, and effectively divides it into large numbers of minute excretory units (the **metanephric tissue caps**) which differentiate to form the **metanephric** (or **nephric**) **vesicles**. The first several orders of branches of the ureteric bud derivatives subsequently amalgamate to form the **renal calyces**.

By about *weeks 10–12*, the excretory units each form functional vesicles or **nephrons**, and each unites with a small branch of the ureteric bud-derived **collecting duct** system. Each excretory unit differentiates to form a **Bowman's capsule** and a **proximal** and **distal convoluted tubule**, and a **loop of Henle**. The urine that is produced passes into the collecting ducts and then drains into the renal calyceal system. It then passes, via the **ureters**, into the **bladder**, where the urine is stored. The bladder is a derivative of the cloacal region of the primitive hindgut, formed by the downgrowth of the urorectal septum. This divides the primitive hindgut into a dorsal 'gut' component and the ventrally located urogenital sinus. It is the latter which differentiates into the bladder. Throughout gestation, it is the urine produced by the fetal kidneys which gives rise to the majority of the **amniotic fluid**.

The internal genital duct component of the urogenital system and the gonads are initially morphologically and histologically identical in the two sexes up to about the end of *week 6*, and for this reason, this period is termed the **'indifferent' stage**. The critical factor involved in inducing the differentiation of the internal genital duct system in the direction of maleness or femaleness, is believed to be related to the genetic sex of the **primordial germ cells**. These migrate from the wall of the **yolk sac** to the **gonadal ridges** during *weeks 4–6*.

In male embryos, which possess an XY sex chromosome constitution, a **testis-determining gene**, *SRY*, is located on the **sex-determining region of the Y chromosome**, and it is through the influence of this factor that the gonad is induced to develop into a **testis**. During *week 7*, the **cortical cords** (or **primitive sex cords**) invade the subjacent mesenchyme in the gonadal region of the **urogenital ridge** and differentiate into the **seminiferous tubules**, and these surround the primordial germ cells which are located in the future medullary region of the testis. **Anti-Müllerian hormone (AMH)** is then produced by the **Sertoli cells**, and this is believed to be responsible for the differentiation of **Leydig** (or **interstitial**) **cells** which subsequently secrete testosterone and other androgenic hormones which influence the development of the external genitalia and other secondary sex features. AMH is believed to have a *positive* influence on the mesonephric duct, inducing it to differentiate into the **efferent ducts**, the **ductus deferens**, the **ejaculatory ducts** and the **seminal vesicles**, while at the same time causing the *regression* of the paramesonephric ducts.

It is presently unclear whether there are sex-determining factors in the female comparable to those known to be active in the male. It has been suggested that it is likely to be the *absence* of the sex-determining factors on the Y chromosome that induces the development of femaleness to occur. In the female, the gonad differentiates into the ovary, while the paramesonephric ducts differentiate into the **uterus, oviducts** and **upper part of the vagina**, the lower part of the vagina being of urogenital sinus origin. The mesonephric ducts almost completely regress in the female, and this is equivalent to the situation observed in the male where the paramesonephric ducts regress.

The histological features of the testis are readily recognizable by about *weeks 10–12*, whereas those of the ovary only develop their characteristic features by about *week 16*.

2.7 Derivatives of the head and neck, and of the pharyngeal (branchial) arch system

Most of the critical events associated with the development of the head and neck region take place between *weeks 4* and *10*. Following the closure of the rostral neuropore, on or about *day 24*, the various elements that give rise to the face (the **facial swellings** or **processes**) fuse together. By about *day 28*, the paired **frontonasal processes**, and a series of externally delineated **pharyngeal** (or **branchial**) **arches** are clearly seen. The first of the pharyngeal arches is the most impressive, and gives rise to the **maxillary** and **mandibular swellings**. The various facial processes surround the buccopharyngeal membrane, and this subsequently breaks down during the second half of *week 3*.

During the first part of *week 5*, the central part of each of the two **nasal placodes**, which form on either side of the frontonasal region, indents to form the **nasal pits**. These then become bounded on their medial and lateral sides respectively by the **medial** and **lateral nasal processes**. The pits indent further, to form the **nasal cavities**, and these become continuous with the nasopharynx during the second half of *week 5*.

The **palatal shelves of the maxillae** differentiate during *weeks 7–9*, and eventually fuse across the midline during *week 10*. At about the same time, the **nasal septum** grows down from the roof of the nasopharynx, and fuses in the midline with the upper surface of the palate, thus separating the oropharynx (below) from the nasopharynx (above), and the latter into the two definitive nasal cavities.

During *week 3*, a diverticulum from the roof of the stomodaeum, termed **Rathke's pouch**, grows upwards towards the floor of the third ventricle, in the region of the diencephalon. The latter, in due course, gives rise to the **posterior pituitary** (or **neurohypophysis**) and the **pituitary stalk** (the **infundibulum**). The Rathke's pouch, which soon becomes detached from the roof of the oral cavity, gives rise to the **adenohypophysis**, the component parts being the **pars anterior**, the **pars intermedia** and the **pars tuberalis**.

In the floor of the oropharynx, during *weeks 3–4*, a series of **lingual swellings** (the primordia of the tongue) form, being derived from each of the first four pharyngeal arches. They subsequently enlarge and eventually fuse together during *weeks 6* and *7*, although the tongue mass continues to increase in volume throughout pregnancy. The **intrinsic and extrinsic muscles of the tongue** (with the exception of palatoglossus) are derived from the **occipital somites**, and consequently receive their nerve supply from the hypoglossal (XII) cranial nerve. The mucous membrane overlying the tongue is derived from the pharyngeal arch endoderm.

The derivatives of the first pharyngeal arch (the two larger **lateral lingual swellings** and the smaller single **median lingual swelling**) give rise to much of the anterior two-thirds of the tongue. General sensation from this region comes via the lingual branch of the mandibular division of the trigeminal (V) cranial nerve, the specific cranial nerve supply to the first pharyngeal arch. The **copula**, the midline swelling derived from the second pharyngeal arch, is completely overgrown at an early stage, principally by the lateral lingual swellings and by the derivative of the third pharyngeal arch. Taste sensation to the anterior two-thirds of the tongue is subserved by the chorda tympani branch of the facial (VII) cranial nerve, the specific cranial nerve supply to the second pharyngeal arch.

A larger swelling, the **hypobranchial eminence**, forms in the floor of the pharynx, and represents the combined derivatives of the third and fourth pharyngeal arches. The posterior one-third of the tongue is principally formed from the derivative of the third pharyngeal arch, and both general and special sensation (i.e. taste) is consequently supplied by the glossopharyngeal (IX) cranial nerve, the specific nerve supply to the third pharyngeal arch. The base of the tongue and epiglottis are derived from the tissue of the fourth pharyngeal arch, and are consequently supplied by the superior laryngeal branch of the vagus (X) cranial nerve.

The **thyroid gland** is the first of the endocrine organs to differentiate, and develops as a downgrowth from the **foramen caecum**, a midline indentation on the dorsum of the tongue located between its anterior two-thirds and posterior one-third. The downgrowth from the foramen caecum, the **thyroglossal duct**, is first seen towards the end of *week 4*. It bifurcates to give rise to the two lobes of the thyroid gland, and these become associated with the laryngeal cartilages during *week 7*. Occasionally, remnants of the thyroglossal duct persist as **thyroglossal cysts**, and its lower part may persist as a **median** or **pyramidal lobe** of the thyroid gland.

Between *weeks 5* and *7*, a series of mostly glandular structures are derived from the various pharyngeal pouches. The first pouch is the exception in this regard, in that it gives rise to the **tubotympanic recess**, and will in due course form the **cavity of the middle ear**, being connected to the oropharynx by the **pharyngotympanic (or Eustachian) tube**.

The **palatine tonsils** arise as outpouchings of the endoderm which lines the second pharyngeal pouch. These penetrate into the subjacent mesenchyme tissue which forms the **tonsillar stroma**. Lymphoid tissue aggregates are not generally seen in this location before about *week 20*, and then constitute a central core (the **tonsillar crypts**) within the endodermal buds.

The **thymus gland** arises from the ventral tubular part of the third pharyngeal pouch. The two thymic primordia enlarge, fuse together across the ventral midline, and then descend towards the superior mediastinum, taking with them the derivatives of the dorsal part of the third pharyngeal pouch, the **parathyroid IIIs** (or **inferior parathyroids**). These are so-named because, in their definitive location, they are normally found embedded in the thyroid gland *below* the parathyroid glands derived from the dorsal part of the fourth pharyngeal pouches (the so-called **parathyroid IVs**, or **superior parathyroids**). The **ultimobranchial bodies** are derived from the ventral tubular part of the fourth pharyngeal pouch. These structures amalgamate with the thyroid gland, and their cells become disseminated within it. They are believed to give rise to the **C-cells** (or **parafollicular cells**) which produce calcitonin, and are involved in calcium homeostasis.

The **lungs and lower respiratory tract** develop from an anteriorly directed diverticulum in the floor of the pharyngeal region of the foregut, termed the **laryngotracheal groove**. This appears on or about *day 22*, and is located just caudal to the fourth pharyngeal arch. It rapidly deepens to give rise to the **tracheal diverticulum**, whose proximal part forms the **trachea**, while the distal part bifurcates to give rise to the two **main bronchi**. The latter are surrounded by splanchnic mesenchyme and form the two **lung buds**. These bulge laterally into the **pericardio-peritoneal canals**, which are lined with coelomic epithelium, and which will subsequently cover the surface of the lungs and line the pleural cavities.

Successive orders of branching of the bronchial tree form the **secondary** and **tertiary bronchi**, and by about *week 16* of gestation **terminal bronchioles** are formed. During the period between *weeks 16* and *24–28*, **respiratory bronchioles** are formed, and these subdivide to give rise to the **primitive alveoli**. From about *week 36*, the alveoli progressively mature.

3. Gross morphological differences between the human and mouse embryo/fetus

While the substantial temporal difference between the development of the human and the mouse which are visible in histological sections has been emphasized in the previous two parts of this chapter, in this section the gross anatomical differences between the two species will be highlighted. As has been indicated in the introductory section, major differences exist between these two species which are particularly evident during the early postimplantation period up to the early limb-bud stage of development (see below), by which time the external configuration of the human and mouse embryos is remarkably similar.

3.1 Species differences in the detailed timing of the critical morphogenetic events that occur during the preimplantation period

While the exact timing of ovulation and implantation vary to some extent among individuals, both of these events are under hormonal control. Folliculogenesis is stimulated by the pituitary-derived gonadotrophin follicle stimulating hormone (FSH), and ovulation occurs at a specific time after the release of luteinizing hormone (LH) from the pituitary (Edwards, 1980). In the human and in most, but by no means all, strains of mice folliculogenesis and ovulation may be induced following exposure to exogenous synthetic gonadotrophins. If successsful fertilization results following mating, then the first cleavage division and shortly afterwards the two-cell stage is achieved. The latter event occurs at a specific time after sperm–egg interaction; this stage is succeeded by a series of cleavage divisions until the eight- or 16-cell stage is achieved. In the mouse, compaction occurs at this time, with the formation of the morula when the individuality of the formerly discrete blastomeres is eventually lost.

In the human, compaction and morula formation usually occur at least one and usually two cleavage divisions later than in the mouse. At whatever stage this occurs, it represents a critical morphogenetic event, and is the first time during embryonic development when one population of cells is completely surrounded by a second population of cells. The fate of these two populations is now said to be determined, as the inner group is destined to form the inner cell mass (ICM), while the outer cells are destined to form trophectoderm (Tarkowski, 1961). This has subsequently been termed the inside-outside hypothesis. Shortly afterwards, the fluid that accumulates between the cells of the morula gradually unites to form a single fluid-filled space termed the blastocoele. The embryo is now at the early blastocyst stage, and usually contains between 32 and 64 cells; an expanded blastocyst may contain well over 100 cells in the mouse, and several hundred cells in the human. At the blastocyst stage, only about one-third of these cells are within the ICM, the remainder forming the trophectoderm shell. More importantly, it is only the ICM-derived cells that will give rise to the embryo proper; the trophectoderm is destined to give rise exclusively to the embryonic/fetal component of the placenta and the extra-embryonic membranes.

3.2 The fate of the ICM during the early postimplantation period

The immediate fate of the ICM in the two species is different. In the advanced blastocyst stage in the human, a distinct layer of cells delaminates from the blastocoelic surface of the ICM to form the embryonic (proximal or visceral) endoderm. At a slightly later stage these cells migrate eventually to line the entire blastocoelic surface of the mural trophectoderm. At about this time, a cavity forms within the ICM, termed the amniotic cavity. The embryonic disc is then seen to consist of two layers (i.e. it is bilaminar), the epiblast (which faces the amniotic cavity and retains its pluripotentiality) and the hypoblast (or endoderm which faces the blastocoelic cavity). In due course, the shell of mural trophectoderm which covers the layer of visceral endoderm becomes separated from the subjacent endoderm by a layer of extra-embryonic mesoderm, while in the region overlying the ICM the polar trophectoderm also becomes separated from the subjacent amniotic ectoderm by a layer of extra-embryonic mesoderm. Of critical importance is the fact that the region of the conceptus which gives rise to the embryo, the embryonic disc, is bilaminar and flat.

In the mouse, the ICM gives rise to the so-called egg cylinder. The latter arrangement is characteristically seen in rodents, and is believed to be an evolutionary adaptation to maximize the efficiency of the limited space available in the uterine horns of rodents to facilitate the development of large numbers of conceptuses. The cells of the embryonic endoderm delaminate as in the human embryo and initially cover the blastocoelic surface of the ICM; subsequently, these cells spread laterally until they cover the entire blastocoelic surface of the mural trophectoderm. A considerable degree of proliferative activity is then seen within the ICM to form a mass of totipotential epiblast cells, which expands in volume and grows towards the abembryonic pole of the implanting blastocyst. It is this mass of totipotential cells, which are covered with a layer of visceral endoderm, that constitutes the egg cylinder. By about day 5.5 p.c., the proamniotic cavity forms within the egg cylinder. The ectoderm within the distal half of the egg cylinder is destined to form the embryo, while the proximal half gives rise to extra-embryonic tissues. By about day 7 p.c., an anterior and a posterior amniotic fold appear, and shortly afterwards amalgamate. At this stage, the advanced egg cylinder becomes subdivided into three compartments, each with its own discrete cavity: the amniotic cavity, the exocoelomic cavity and the ectoplacental cavity, respectively, distally to proximally. By this stage, the polar trophectoderm proliferates to form the ectoplacental cone, the precursor of the embryonic component of the placenta. The differentiation of the egg cylinder to form the primitive streak stage in the mouse is illustrated diagrammatically (*Figure 2*).

The primitive streak also appears on about day 7 p.c. in the mouse, and shortly afterwards the process of gastrulation is initiated, with the first appearance of intra-embryonic mesoderm. It is important to note that at this stage the configuration of the mouse embryo is such that the embryonic 'disc' (now trilaminar, because its ectoderm layer is separated from the subjacent layer of endoderm by the intervening mesoderm) is curved rather than flat as in the human embryo at the equivalent stage of embryogenesis. More particularly, its dorsal surface is concave, while its ventral surface is convex. Since the dorsal surface gives rise to the neural axis and surface ectoderm, while the ventral surface is exclusively associated with the differentiation of the gut, the germ layers are said to be 'inverted' because the embryo, if it were to continue to develop in this configuration, would be 'inside-out' compared to the human embryo at a comparable stage of development. In order to adopt the human configuration, the mouse embryo undergoes the process of 'turning' (or axial rotation).

3.3 The process of 'turning' by which the mouse embryo adopts the characteristic 'fetal' position, also allows it to become surrounded by its extra-embryonic membranes

Once the mouse embryo possesses about 6–8 pairs of somites, it initiates the process of 'turning'. This is a complex process which is also best described diagrammatically (*Figure 3*). Essentially, over the 24-h period between days 8.5 and 9.5 p.c., a change is brought about in the configuration in the embryo so that at the end of this period the dorsal surface of the embryo becomes convexly curved, while its ventral surface, initially convex, becomes concave with respect to the embryonic axis. By this manoeuvre, the mouse embryo adopts the characteristic so-called 'fetal' position. At the same time as the embryo 'turns', it also rolls into and becomes surrounded by its extra-embryonic membranes; the amnion, within which the amniotic fluid is located surrounds and bathes the embryo, and this in turn is surrounded by the visceral yolk sac (for further details of the 'turning' process, see Kaufman, 1992).

In both human and mouse, the mesodermal component of the yolk sac is the

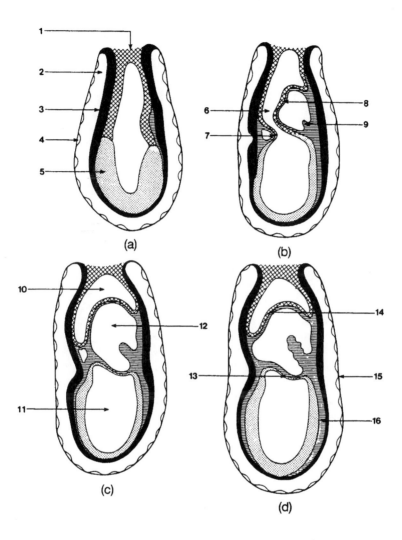

Key:
1. Ectoplacental cone
2. Yolk sac cavity
3. Visceral endoderm
4. Parietal endoderm
5. Embryonic ectoderm
7. Anterior amniotic fold
8. Posterior amniotic fold
9. Allantois
10. Ectoplacental cavity
11. Amniotic cavity
13. Amnion
14. Chorion
15. Reichert's membrane
16. Mesoderm

Figure 2. Stages in the conversion of the egg cylinder into the primitive-streak-stage mouse embryo: (a) egg cylinder stage; (b) proamniotic canal present; (c) during closure of the proamniotic canal; (d) after closure of the proamniotic canal. Reproduced from Kaufman (1990) by permission of Oxford University Press.

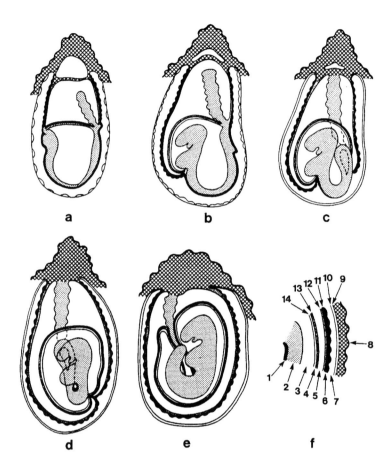

Key to (f):

1. Embryonic endoderm
2. Embryonic ectoderm + mesoderm
3. Amniotic cavity
4. Amnion
5. Exocoelomic cavity
6. Yolk sac
7. Yolk sac cavity

8. Ectoplacental cone and trophectoderm derivatives
9. Reichert's membrane
10. Parietal (extra-embryonic) endoderm
11. Visceral (extra-embryonic) endoderm
12. Extra-embryonic mesodermal component of yolk sac
13. Mesodermal component of amnion
14. Ectodermal component of amnion

Figure 3. Simplified diagrammatic sequence to illustrate changes in the conformation of the mouse embryo and the way in which the extra-embryonic membranes surround it as it undergoes the process of 'turning': (a) presomite headfold stage, 7.5–8 days; (b) embryo with 8–10 pairs of somites, about 8.5 days; (c) embryo with 10–12 pairs of somites, about 8.75 days; (d) embryo with 12–14 pairs of somites, about 9 days; (e) embryo with 15–20 pairs of somites, about 9.5 days; (f) embryonic layers, extra-embryonic tissues and cavities encountered in the embryo illustrated in (e). Reproduced from Kaufman (1990) by permission of Oxford University Press.

source of the haemopoietic stem cell lineage and is probably also the initial site of differentiation of the primordial germ cells. In the human embryo, that portion of the secondary yolk sac that becomes enclosed within the embryo becomes subdivided into the three components of the primitive gut: the foregut, midgut and hindgut. Only the ventral part of the midgut region of the primitive gut communicates with the definitive yolk sac (or vitelline sac) via the vitello-intestinal duct (located in the body stalk, which subsequently differentiates into the umbilical cord). With the subsequent growth of the embryo, the length of the duct increases, and the definitive yolk sac becomes increasingly distant from the body wall, being located at the placental end of the umbilical cord. What little remains of the definitive yolk sac regresses and usually completely disappears between the fifth month and full term. In the mouse, the extra-embryonic vascular system which develops within the wall of the yolk sac plays an important role in transporting nutrients across to the conceptus, and is retained until the end of gestation. ˙

3.4 Placentation and configuration of the extra-embryonic membranes

The amnion is the first of the extra-embryonic membranes to form. In the human, this membrane consists of a bilayer of flattened epithelial cells which are continuous on all sides with the margins of the embryonic disc; it consequently forms the boundary of the amniotic cavity, and first completely surrounds the embryo at about the early limb bud stage. In the mouse, the amnion is formed following the fusion of the anterior and posterior amniotic folds and the closure of the pro-amniotic canal at the early primitive streak stage; it is also located at the peripheral margin of the embryonic disc. However, the amnion only completely surrounds the embryo after the completion of the 'turning' process (*Figure 2*).

Once the embryonic axis has been established, at the primitive streak stage, the process of embryogenesis is initiated, and very shortly afterwards is associated with the establishment of the three definitive germ layers: the ectoderm, endoderm and mesoderm. While there are significant differences in the configuration of the human and mouse conceptuses at this time (see above), this is even more evident when a comparison is made between the type of placentation observed and fate of the yolk sac in these two species: in the human the yolk sac, after playing a critical initial role in haematopoiesis, regresses, but in the mouse it is retained and plays an important part in facilitating the transfer of nutrients to the conceptus. While a general consideration of placentation is beyond the remit of this chapter, the interested reader should refer for additional information to Amoroso (1952), Hamilton and Mossman (1972) and Steven (1975).

3.5 The venous drainage of the head, neck and forelimbs/upper limbs

The persisence of the primitive symmetrical arrangement whereby both the left and right superior vena cavas are retained in the mouse is quite different from the normal arrangement seen in the human, where both sides of the head and neck as well as both upper limbs drain into the single (right-sided) superior vena cava. The primitive symmetrical arrangement is modified in the human during about the eighth week p.c. with the establishment of the left brachiocephalic (innominate) vein. This represents a large anastomotic channel which forms between the left and right anterior cardinal veins. Concomitant with the appearance of this new vessel, the proximal part of the left anterior cardinal vein where it communicates with the left common cardinal vein disappears. In the mouse, by contrast, the left superior vena cava is retained and unites with the proximal part of the inferior vena cava. The venous blood from the left side of the head and neck and the left forelimb thus drains into the *inferior* part of the right atrium. This arrangement means that the dynamics of the blood flow into the right atrium during the fetal period in these two species differs markedly. In both species, however, the majority of the left posterior cardinal venous system disappears, only the right inferior vena cava being normally retained. Accordingly, in both species, the venous drainage from the left side of the body caudal to the diaphragm is directed towards the right side.

3.6 The parathyroid glands

In the mouse, only a single pair of parathyroid glands develops. By contrast, in the human, two pairs of parathyroid glands differentiate: one pair from the dorsal part of the third and a second pair from the dorsal part of the fourth pharyngeal pouches. These are termed the parathyroid III glands, or inferior parathyroids, and the parathyroid IV glands, or superior parathyroids, respectively, because of their pharyngeal arch origin and because of their definitive location in relation to the thyroid gland. In rodents, parathyroid glands only develop from the derivatives of the third pharyngeal pouches.

3.7 The lobes of the lungs

The arrangement of the subdivisions of the lungs in the two species is quite different. While both possess both a left and right lung, the number of lobes present in each is different. The human left lung is usually subdivided into two lobes, while the right lung is usually subdivided into three lobes. By contrast, the left lung of the mouse is retained as a single unit, whereas the right lung is subdivided into four lobes, the cranial, middle and caudal lobes as well as the rather anomalous accessory lobe. The latter, while undoubtedly a component of the right lung as evidenced by the origin of its main bronchus, crosses the midline and is principally located in the left side of the thoracic cavity.

3.8 Intra-abdominal viscera

While both species possess the expected range of intra-abdominal organs that, in developmental terms, are formed as outgrowths from the primitive gut, such as the liver and pancreas, their gross morphology is dissimilar in the two species, and probably reflects differences in their dietary habits. The pancreas in particular appears to be a much more diffuse organ in the mouse than in the human, and its secretions drain into the duodenum via a substantial number of separate ducts rather than via a single (the 'main') pancreatic duct or occasionally via two ducts (the 'main' accompanied by an 'accessory' duct). While in both species the liver is derived from left and right components, the detailed anatomical arrangement is somewhat dissimilar, but a detailed account of these differences is beyond the scope of this chapter.

Due to the large volume of the liver and to a lesser extent, that of the stomach and mesonephroi on or about day 12 p.c. in the mouse, relatively little space is available within the abdominal cavity for the gut to elongate and differentiate. It is for this reason that the midgut loop is extruded into the so-called 'physiological' umbilical hernia. Only when adequate space becomes available within the abdominal cavity, with the regression of the mesonephroi and their replacement by the much smaller metanephroi (the definitive kidneys), and due to the proportionately decreased volume occupied by the liver and stomach at this time, can the midgut loop return to the abdominal cavity and the discontinuity in the anterior abdominal wall in the region of the umbilicus finally close; this occurs during day 16 p.c. In the human, a physiological umbilical hernia also develops, and for the same reason. This is first seen at about eight weeks p.c.; the midgut loop begins to return to the abdominal cavity during week 10, and the process is completed during week 12 p.c.

The mouse and rat are, in many ways, remarkably similar, but some important anatomical differences are observed between these two species. While the mouse, like the human, possesses a gall bladder, the biliary system in the rat is exclusively intrahepatic.

3.9 The female internal genital duct system

Of the intra-abdominal viscera, the internal genital duct system of the female mouse, while having a similar embryological origin to its human equivalent, is in gross anatomical terms markedly different. This is principally because the mouse retains the primitive arrangement whereby the paired paramesonephric ducts remain as separate entities along the majority of their length, being only united in their most proximal part, the region of the future cervix and upper part of the vagina. As a consequence of the latter arrangement, two distinct uterine horns develop, each with their own more distally located oviduct (or Fallopian tube). In the human, the characteristic primate arrangement is seen, whereby the paired paramesonephric ducts unite along the majority of their length, giving rise to a single midline uterine body and (proximally located) cervix with a pair of laterally (and distally) located oviducts.

A pair of coagulating glands are present in the male mouse. These are located close to the seminal vesicles, and their ducts open into the prostatic urethra. Their secretions pass into the ejaculate and cause it to coagulate shortly after the latter enters the vagina. A white crystalline structure forms which fills the vagina, and is, for obvious reasons usually termed the vaginal 'plug'. This conveniently provides evidence of a successful mating. A similar arrangement is also seen in the rat, but the 'plug' tends to be a more transient structure, and often drops out of the vagina after only a few hours, rather than persisting *in situ* for up to 12 or more hours, as is commonly the case in mice.

3.10 Species differences in the number of vertebrae present in the different regions of the vertebral axis

With regard to the skeleton, the most obvious species difference, apart from the proportionate lengths of the various components of the appendicular skeleton, relates to the number of vertebral elements that are associated with each anatomical subdivision of the vertebral column. In the human, there are usually seven cervical, 12 thoracic (or dorsal), five lumbar and five sacral vertebral units (the latter fuse together to form the sacrum), and four coccygeal vertebrae; initially about eight to ten coccygeal elements are present, although the most caudal of these usually fuse together to form a single unit (the coccyx). Occasionally, the most caudal of the cervical vertebrae may have typical thoracic features (with associated rib elements), and the most caudal of the lumbar vertebrae may be partly or completely fused with the most rostral part of the sacrum (termed sacralization). Conversely, the most rostral part of the sacrum may be present as a distinct vertebra which accordingly possesses lumbar-like features (termed lumbarization). More rarely, the first cervical vertebra may be partly or completely fused to the occipital region of the skull (termed occipitalization). In certain pathological conditions (e.g. the Klippel–Feil sequence), numbers of cervical vertebrae may fuse together (for review of syndromes with vertebral defects, see Jones, 1988). In the mouse, the usual number of vertebral units associated with each region of the vertebral column is as follows: seven cervical; 13 thoracic; six lumbar; four sacral; 27–31 coccygeal, which comprise the vertebrae of the tail; the maximum number of pairs of somites, from which the vertebral units will form, are usually present in the mouse by about day 13.5 p.c. Occasionally there may be variations in the number of vertebrae present in this species, particularly in the thoracic and cervical regions, and this feature is usually strain dependent (Grüneberg, 1963).

During early human development, the tail is an obvious feature, but gradually regresses, so that by the end of the second month, in specimens of about 30 mm crown–rump length, the tail has almost completely disappeared. In tailless (T/t) mouse mutants, and in other mutants in which abnormalities of the tail are a characteristic feature, this condition is usually associated with either gross abnormalities of the notochord (in T/t embryos the notochord in the tail region is incorporated into the neural tube) or the latter may be completely deficient in the tail region, as occurs in Danforth's short-tail (Sd/Sd), truncate (tc/tc) and pintail $(Pt/+$ and $Pt/Pt)$ mutants; the data from these mutants strongly suggests that for the formation of caudal vertebrae in the mouse, the presence of a (functioning) notochord is essential (for review, see Grüneberg, 1963).

3.11 Other species differences

Although of only academic interest, it should be noted that the number of vallate papillae (the large dome-shaped taste receptors) present on the dorsum of the tongue at the boundary between the anterior two-thirds and the posterior one-third varies considerably even between rodents. In the mouse and rat, a single large median circumvallate papilla is present. This contrasts with the situation in the Japanese dormouse which possesses three, and the porcupine and anteater which have two (for references, see AhPin *et al.*, 1989). In the human, between eight and 12 may be present in the form of a V-shaped row immediately in front of and parallel with the sulcus terminalis. In all species, the vallate papillae are supplied by the glossopharyngeal (IX) cranial nerve. In the mouse, the single midline circumvallate papilla has a bilateral innervation from this source.

References

Adinolfi M, Polani PE. (1989) Prenatal diagnosis of genetic disorders in pre-implantation embryos: invasive and non-invasive approaches. *Hum. Genet.* **83:** 16–19.

AhPin P, Ellis S, Arnott C, Kaufman MH. (1989) Prenatal development and innervation of the circumvallate papilla in the mouse. *J. Anat.* **162:** 33–42.

Amoroso EC. (1952) Placentation. In: *Marshall's Physiology of Reproduction. Vol. 2,* 3rd Edn. (ed. AS Parkes), Longmans Green, London, pp. 127–311.

Balinsky BI. (1981) *An Introduction to Embryology,* 5th Edn. Holt, Rinehart & Winston, New York, NY.

Bard JBL. (1990) *Morphogenesis. The Cellular and Molecular Processes of Developmental Anatomy.* Cambridge University Press, Cambridge.

Bard JBL, Kaufman MH. (1994) The mouse. In: *Embryos: Color Atlas of Development* (ed. JBL Bard). Wolfe Publishing, London, pp. 183–206.

Bergsma D. (1979) *Birth Defects Compendium.* 2nd Edn. Macmillan, London.

Boué J, Boué A. (1973) Anomalies chromosomiques dans les avortements spontanés. In: *Les Accidents Chromosomiques de la Reproduction.* (eds A Boué, C Thibault), Institut National de la Santé et de la Recherche Médicale, Paris, pp. 29–55.

Butler H, Juurlink BHJ. (1987) *An Atlas for Staging Mammalian and Chick Embryos.* CRC Press, Boca Raton, FL.

Copp AJ, Cockroft DL. (1990) *Postimplantation Mammalian Embryos. A Practical Approach.* IRL Press, Oxford.

Edwards RG. (1980) *Conception in the Human Female.* Academic Press, London.

Feichtinger W, Kemeter P. (1987) *Future Aspects In Human In Vitro Fertilization.* Springer-Verlag, Berlin.

Findlater GS, McDougall RD, Kaufman MH. (1993) Eyelid development, fusion and subsequent reopening in the mouse. *J. Anat.* **183:** 121–129.

Firth HV, Boyd PA, Chamberlain P, MacKenzie IZ, Lindenbaum RH, Huson SM. (1991) Severe limb abnormalities after chorion villus sampling at 56–66 days' gestation. *Lancet* **337:** 762–763.

Froster-Iskenius UG, Baird PA. (1989) Limb reduction defects in over one million consecutive livebirths. *Teratology* **39:** 127–135.

Gardner RL, Edwards RG. (1968) Control of the sex ratio at full term in the rabbit by transferring sexed blastocysts. *Nature* **218:** 346–348.

Gasser R. (1975) *Atlas of Human Embryos.* Harper & Row, Hagerstown, MD.

Gilbert SF. (1994) *Developmental Biology,* 4th Edn. Sinauer Associates Inc., Sunderland, MA.

Grüneberg H. (1963) *The Pathology of Development. A Study of Inherited Skeletal Disorders in Animals.* Blackwell Scientific Publications, Oxford.

Hamburger V, Hamilton HL. (1951) A series of normal stages in the development of the chick embryo. *J. Morph.* **88:** 49–92.

Hamilton WJ, Mossman HW. (1972) *Hamilton, Boyd and Mossman's Human Embryology. Prenatal Development of Form and Function,* 4th Edn. W. Heffer & Sons Ltd, Cambridge.

Hebel R, Stromberg MW. (1986) *Anatomy and Embryology of the Laboratory Rat.* BioMed Verlag, Worthsee.

Hertig AT, Rock J, Adams EC. (1956) A description of 34 human ova within the first 17 days of development. *Am. J. Anat.* **98:** 435–493.

Hogan B, Beddington R, Constantini F, Lacy E. (1994) *Manipulating the Mouse Embryo. A Laboratory Manual,* 2nd Edn. Cold Spring Harbor Laboratory, New York, NY.

Jones KL. (1988) *Smith's Recognizable Patterns of Human Malformations,* 4th Edn. WB Saunders, Philadelphia, PA.

Kalousek DK, Fitch N, Paradice BA. (1990) *Pathology of the Human Embryo and Previable Fetus: An Atlas.* Springer-Verlag, New York, NY.

Kaufman MH. (1990) Morphological stages of postimplantation embryonic development. In: *Postimplantation Mammalian Embryos: A Practical Approach* (eds AJ Copp, DL Cockroft). IRL Press, Oxford, pp. 81–91.

Kaufman MH. (1992) *The Atlas of Mouse Development.* Academic Press, London.

Kaufman MH. (1994) Hypothesis: the pathogenesis of the birth defects reported in *cvs*-exposed infants. *Teratology* **50:** 377–378.

Keibel F. (1937) *Normentafl zur Entwicklungsgeschichte der Wanderratte (Rattus norvegicus Erxleben).* Fischer, Jena.

Larsen WJ. (1993) *Human Embryology.* Churchill Livingstone, New York, NY.

Moore KL, Persaud TVN. (1993) *The Developing Human*, 4th Edn. WB Saunders Company, Philadelphia, PA.

O'Rahilly R, Müller F. (1987) *Developmental Stages in Human Embryos.* Carnegie Institute Publication No. 637, Washington, DC.

Otis EM, Brent R. (1954) Equivalent ages in mouse and human embryos. *Anat. Rec.* **120**: 33–64.

Pearson AA. (1980) The development of the eyelids. Part I. External features. *J. Anat.* **130**: 33–42.

Rugh R. (1968) *The Mouse. Its Reproduction and Development.* Burgess Publishing Company, Minneapolis.

Rugh R. (1990) *The Mouse. Its Reproduction and Development.* Oxford University Press, Oxford.

Sabbagha RE. (1987) *Diagnostic Ultrasound Applied to Obstetrics and Gynecology*, 2nd Edn. J.B. Lippincott Company, Philadelphia, PA.

Schloo R, Miny P, Holzgreve W, Horst J, Lenz W. (1992) Distal limb deficiency following chorionic villus sampling? *Am. J. Med. Genet.* **42**: 404–413.

Seppälä M, Edwards RG. (1985) In vitro fertilization and embryo transfer. *Ann. N.Y. Acad. Sci.* **442**: 1–619.

Singer P, Kuhse H, Buckle S, Dawson K, Kasimba P. (1990) *Embryo Experimentation.* Cambridge University Press, Cambridge.

Snell GD, Stevens LC. (1966) Early embryology. In: *Biology of the Laboratory Mouse*, 2nd Edn. (ed. EL Green). McGraw-Hill, New York, pp. 205–245.

Steven DH. (1975) *Comparative Placentation.* Academic Press, London.

Tarkowski AK. (1961) Mouse chimaeras developed from fused eggs. *Nature* **190**: 857–860.

Theiler K. (1972) *The House Mouse: Development and Normal Stages from Fertilization to 4 weeks of Age.* Springer-Verlag, Berlin.

Theiler K. (1989) *The House Mouse: Atlas of Embryonic Development.* Springer-Verlag, New York, NY.

Therkelsen AJ, Grunnet N, Hjort T, Jensen OM, Jonasson J, Lauritsen JG, Lindsten J, Petersen GB. (1973) Studies on spontaneous abortion. In: *Les Accidents Chromosomiques de la Reproduction.* (eds A Boué, C Thibault). Institut National de la Santé et de la Recherche Médicale, Paris, pp. 81–93.

Thomson D'AW. (1917) *On Growth and Form.* Cambridge University Press, Cambridge.

Waddington CH. (1956) *Principles of Embryology.* Allen and Unwin, London.

Warburton D, Byrne J, Canki N. (1991) *Chromosome Anomalies and Prenatal Development: An Atlas.* Oxford University Press, New York, NY.

Warkany J. (1971) *Congenital Malformations: Notes and Comments.* Year Book Medical Publishers Inc., Chicago, IL.

Weiss P. (1939) *Principles of Development.* Holt, Rinehart & Winston, New York, NY.

Wilson JG, Fraser FC. (1977) *Handbook of Teratology, Vols. 1–3.* Plenum Press, New York, NY.

Winter RM, Knowles SAS, Bieber FR, Baraitser M. (1988) *The Malformed Fetus and Still Birth: A Diagnostic Approach.* John Wiley & Sons, New York, NY.

Witschi E. (1962) Development: rat. In: *Growth Including Reproduction and Morphological Development* (eds PL Altman, DS Dittmer). Biological Handbooks of the Federation of American Societies for Experimental Biology, Washington, DC, pp. 304–314.

Yanagimachi R, Yanagimachi H, Rogers BJ. (1976) The use of zona-free animal ova as a test-system for the assessment of the fertilizing capacity of human spermatozoa. *Biol. Reprod.* **15**: 471–476.

7

Approaches used to study gene expression in early human development

David I. Wilson

1. Introduction

The use of embryonic tissue from any species for experimental purposes has to be justified both in ethical and scientific terms, and these concerns are of particular importance when studying human embryonic tissue. The use of human embryos for gene expression studies is not universally permissible and in those countries which permit it, prior approval has usually to be obtained from either local or national ethical committees (for review see Burn and Strachan, 1995). Scientists investigating human embryonic gene expression should perhaps consider ethical justification of their work as much a part of the experimental methodology as practical aspects of the experimental techniques and should not divorce themselves from ethical discussion (for further discussion see Chapter 4). Legislation exists in most countries which determines whether human embryos may be used for experimental purposes and this reflects cultural, political and religious concerns (for review see Burn and Strachan, 1995). Within the UK, the Polkinghorne report (Polkinghorne, 1989) has recommended conditions for the use of human embryos in scientific research. It is important that those performing the research are familiar with the legislation and any subsequent changes that occur.

The aim of this chapter is to provide an outline of the different approaches which can be used to analyse human gene expression and to consider their application to the study of gene expression in early human development. In particular, the technique of *in situ* hybridization is discussed in greater depth; although the actual technique is reasonably well established and detailed protocols are well described (Akhurst, 1992; Moorman *et al.*, 1993; Wilkinson, 1992a), its use for human embryo expression is relatively novel. It is not the aim of this review to

Molecular Genetics of Early Human Development, edited by T. Strachan, S. Lindsay and D.I. Wilson.

describe laboratory *in situ* protocols, but for the unfamiliar reader the general principles of tissue *in situ* hybridization will be presented and in particular those aspects that are particularly important when investigating human embryos.

2. Methods used to study gene expression

Many techniques have been developed to analyse gene expression within cells and tissues, whether eukaryotic or prokaryotic, embryonic or adult. Genes are transcribed into mRNA and in the vast majority of cases, the resultant RNA sequences are translated into polypeptides. A variety of approaches are available for assaying expression at both the RNA and polypeptide levels and these can be divided into two general groups:

(i) Those where RNA or protein is extracted from a source tissue and is subsequently fractionated in some way.
(ii) Those that detect RNA or protein within intact cells.

To a certain extent the two sets of approaches are complementary; the methods used to study extracted expression products are rapid but inevitably the expression profiles are much less complex than the very high resolution patterns obtained using the rather laborious methods which survey gene expression in intact cells. In the latter case there is limited opportunity for surveying the expression of many genes at a time, but some of the methods used to study extracted expression products can simultaneously track the expression of numerous genes, and potentially all the genes, in the genome of interest.

2.1 Gene expression methods which assay extracted RNA/cDNA or protein

Methods of this type involve extracting and purifying RNA or protein from the tissue or cell source, or converting RNA to cDNA using reverse transcriptase (RT) and amplifying expressed sequences, either by RT–PCR (polymerase chain reaction) or by cloning the DNA into bacterial cells to generate cDNA libraries. They include some well established methods, which have a fairly limited capacity for surveying gene expression plus a variety of recently developed methods, which allow tracking of multiple genes simultaneously.

Methods with a limited expression screening capacity. Some well-established methods are included in those with limited expression screening capacity, such as Northern and RNA dot-blot hybridization, ribonuclease protection assays, RT-PCR and Western blotting. Such techniques are generally rapid, straightforward and do not require expensive equipment. They provide valuable information on the relative abundance of expression products and, in many cases information is provided on product sizes, following fractionation on appropriate gel electrophoresis systems. Generally, however, the methods are designed to track the expression of one or a few genes at a time and so they have a limited expression screening capacity. In addition, these methods cannot discriminate between the individual cells or groups of cells within a source tissue and so have limited cellular resolution, unlike the methods used to study gene expression in intact cells (see below).

Methods of surveying expression in many genes at a time. Methods designed to survey the expression of multiple genes simultaneously have mostly been developed very recently. One striking exception is 2-D gel electrophoresis, which allows expression screening at the protein level for many genes at a time and has a long history (Anderson and Anderson, 1996). This technique has been used to study gene expression in a variety of human tissues and several electronic databases have already been developed to record the observed expression patterns (Bodymap). More recently, mass spectroscopy methods have also shown considerable potential for screening protein products. Total protein fractions can be separated chromatographically, individual fractions cleaved and the cleaved peptides analysed by mass spectroscopy (see Chait, 1996 and below).

At the transcriptional level, many novel, high capacity expression screening approaches have recently been devised. Sequencing of cDNA clones in human cDNA libraries was not intended as an expression screen per se (Boguski and Schuler, 1995; Schuler *et al.*, 1996), but it has provided some insights into the relative transcriptional activities of certain genes (Bodymap). Recently, a variety of powerful new methods have been designed to screen the transcriptional activities of multiple genes at a time. Differential display is a PCR-based technique which uses short arbitrary primers to amplify random subsets of the transcripts of tissues (which may be as small as single cells) which are then displayed by size-fractionation on a gel (Laing and Pardee, 1992). A comparison can be made between mRNA from different source tissues, and so differential expression patterns can be identified. However, the products are a subgroup of the total tissue mRNA, having been amplified in an unselected and arbitrary way. Serial analysis of gene expression (SAGE) is a similar method, which rescues very short (9-bp) gene-specific sequence tags from 3' end fragments in bulk cDNA, and identifies the expressed genes by concatenation of individual tags followed by sequencing (Adams, 1996; Velculescu *et al.*, 1995). Finally, multiplex hybridization assays have been devised whereby total cDNA from a cell is labelled with a fluorescent tag and used as a hybridization probe to interrogate micro-arrays of cDNA clones which are robotically spotted onto glass wafers, or of gene-specific oligonucleotides which are synthesized *in situ* on the glass wafers (Anonymous, 1996; DeRisi *et al.*,1996; Lockhart *et al.*, 1996; Schena, 1996; Schena *et al.*, 1995, 1996). Fluorescence detection systems are then able to quantify the hybridization signals, and thence the relative transcriptional activities of multiple genes.

The power of the above methods is inevitably related to their capacity for expression screening, and when the Human Genome Project eventually delivers a complete catalogue of human genes, there will be the potential for *whole genome expression screens*, that is assays which simultaneously monitor the expression of each of our 100 000 or so genes. The known identities of the genes and inferred protein products will inevitably aid the expression screens. At the level of 2-D gel electrophoresis, for example, individual proteins may be identified by comparing observed and expected radiolabelling patterns following incorporation of specific amino acids labelled with different radioisotopes (Maillet *et al.*, 1996). The application of this general approach to studying human proteins is, however, handicapped by technical limitations: the method is difficult to standardize and automate and its capacity for expression screening is limited. Mass spectroscopy approaches, in combination with database searching may offer greater potential (Chait, 1996), but it is still difficult to imagine how improvements in current technology will permit whole genome expression screening at the protein level.

Transcriptional profiling may be an easier route to whole genome expression screening, although some of the techniques described before are more suited than others. Differential display and SAGE, for instance, are useful when novel genomes are being investigated, as they have the capacity to identify new expressed sequence tags (ESTs) or genes. If the genome under investigation has been fully sequenced and all transcripts indentified (as one hopes will be the case for the human genome), these methods are less attractive. Instead, automated micro-array analyses are likely to become the 'gold standard' for simultaneous screening. Already, researchers at Affymetrix have been able to design oligonucleotide micro-arrays that permit simultaneous expression screening of 50 000 human genes, many of unknown function (D. Lockhart, personal communication), and further extension to include the remaining human genes is eagerly anticipated. However, whilst such analysis will be immensely useful as a 'first pass' expression screen, none of the above techniques will provide high resolution expression patterns of the type that can be obtained by *in situ* gene expression studies.

3. *In situ* methods

In situ gene expression studies attempt to retain the spatial distribution of gene products in tissues and cells. Transcriptional products (mRNA) may be analysed by tissue *in situ* hybridization using labelled oligonucleotides or antisense RNA probes (see Section 4). *In situ* hybridization was developed in 1969 by Pardue and Gall and independently by Jones (John *et al.*, 1969; Pardue and Gall, 1969) using radioisotopic labelling of nucleic acids which could be purified by conventional biochemical methods rather than cloned sequences. With the advent of non-radioactive detection methods and advances in molecular biology (e.g. cloning, *in vitro* transcription and labelling techniques), other nucleic acid sequences are now able to be detected. An example of tissue *in situ* hybridization is shown in *Figure 1* and demonstrates expression of TBX5 in human embryos (Yi *et al.*, 1997).

Translational products (protein) can also be detected using poly- or monoclonal antibodies to specific protein epitopes (*in situ* immunocytochemistry)

Figure 1. Tissue *in situ* hybridization. Both sections are hybridized with an ^{35}S-labelled antisense RNA probe for TBX5 and photographed under dark-field illumination. (a) Carnegie Stage 15 human embryo: transverse section through the heart. Expression is seen in the atrial wall, mesoderm surrounding the trachea and thoracic wall. (b) Carnegie Stage 17 human embryo: transverse section at the midthoracic level. Expression is highest in the forelimb and atrial walls. A: atria; V: ventricle; FL: forelimb; C: endocardial cushion.

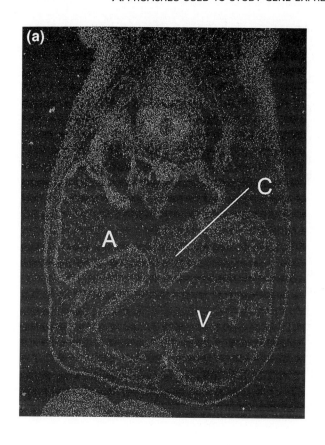

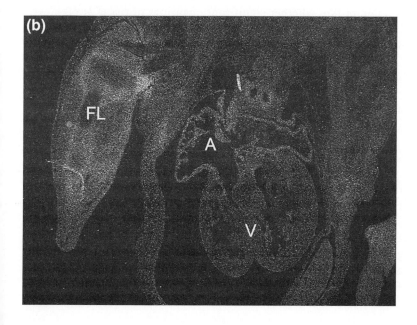

(Christensen and Strange, 1987). Alternatively 'ultra-high resolution' studies may be performed using electron microscopy (EM) and this has allowed subcellular localization of gene products (Binder, 1992). This is a more complex procedure than standard *in situ* or immunocytochemical detection and requires specific tissue fixation and ultrathin sectioning. Preservation of tissue or cellular morphology is paramount and this competes with the conditions that favour hybridization. Nevertheless EM detection, may be required to answer questions about gene interactions at an ultrastructural level.

There are numerous advantages with *in situ* studies of gene expression. Using specific oligonucleotides or antibodies, it is possible to track individual isoforms and determine individual temporospatial expression patterns. The localization of gene products to individual cells is also possible and comparison of mRNA *in situ* hybridization with '*in situ* immunocytochemistry' may determine whether the translated protein co-localizes with the cells that are transcribing the gene. Such high resolution studies are, however, very labour intensive, laborious and time consuming.

Single embryo whole mount *in situ* hybridization has been used to define expression patterns in single whole embryos of *Drosophila*, zebrafish and mice (e.g. see Tautz *et al.*, 1992; Wilkinson, 1992b). This is a powerful approach and, although limited by the physical size of the embryo, may define all the expression domains in a single experiment. This is an advantage over sectioned embryonic tissue where many sections have to be examined to ensure that all expression domains have been defined; this is particularly so if a novel gene is being investigated.

3.1 Approaches for studying gene expression in early human development

It is important to consider which of these methods is suitable for studying gene expression in human embryos. Methods differ as to whether RNA and protein are extracted or remain *in situ*, whether significant quantities of tissue are required and whether many or a few genes can be studied simultaneously. Perhaps the most important factor in deciding which techniques to use, and this will invariably be the limiting factor, is the scarcity of material. All of the previously mentioned approaches can be used if the quantity of embryonic material is 'unlimited'. Clearly this is not the case for human embryonic tissue and complementary approaches are required, thus a combination of *in situ* expression and indirect analyses should be used.

Preimplantation embryos and fertilized oocytes. As discussed elsewhere in this volume, embryos up to 14 days postconception which have been obtained from *in vitro* fertilization (IVF) programmes can, in some cases, be investigated (see Chapter 8). These embryos are not only scarce but contain small numbers of cells and minute quantities of gene products. *In situ* studies would probably be inappropriate, either as whole mount or sectioned material, as the cells and structure are poorly differentiated. Investigations have been directed to those methods that require RNA isolation and include a process of amplification, to provide sufficient material to analyse such as RT-PCR. This can permit assays of even single cells, such as fertilized oocytes, but may be limited in the number of genes which can be analysed for each sample. Also samples can potentially be contaminated by as little as a single cell. It is nevertheless a useful technique. Efforts are also being made to generate cDNA libraries from such embryos (e.g. see Chapter 9). The clones in these libraries may eventually be sequenced which will give information about expression patterning and possibly permit the identification of new genes, particularly those which are so stage-specific in expression that they could otherwise escape detection. An example of this is illustrated in the libraries made by Rosa Beddington and her colleagues, from separate germ layers of early mouse embryos. Tissue from several hundred embryonic mid-gastrulation mouse embryos were dissected and cDNA clone analysis revealed some transcripts to be novel: they had not been identified from other later embryos or adult tissues (Harrison *et al.*, 1995). Currently it seems unlikely that sufficient human material would be available for an equivalent human study, but perhaps the scientific value could justify storing very early human embryos less than 14 days postfertilization, in order to create such libraries, particularly if the cDNA clones were deposited on micro-arrays for screening.

The use of preimplantation embryos for protein analysis is difficult because of the small quantity of tissue and, unlike nucleic acids, amplification of polypeptides or proteins is not possible. However, sensitive enzyme assays and immunocytochemical approaches are possible and provide much information (see Chapter 8).

Studying gene expression in embryos and fetuses. The majority of human embryonic or fetal tissue used in the UK for gene expression studies is obtained from women undergoing termination of pregnancy in the first and second trimesters. This determines that only specific developmental windows can be studied, as most women attend for termination between 4 and 9 weeks postconception. The human developmental period between 14 and 21 days is difficult to study, as it is between the maximum time permitted for *in vitro* culture of IVF embryos (UK) and the earliest time when terminations are performed.

Embryos during this period (4–9 weeks postconception) undergo a rapid increase in size (Chapter 1) and the quantity of tissue from the latter stages is significantly greater than preimplantation embryos or indeed a Carnegie Stage 10 embryo (22 days: 2 mm length). The size at 4 weeks still limits the quantity of mRNA or protein available but the larger embryos, for instance a Carnegie Stage 23 embryo (56 days: 3 cm length), may not restrict the methods that can be used. Nevertheless, embryos from Carnegie Stage 10+ have been used for both indirect extraction methods and *in situ* expression studies. Standard methods, such as RT–PCR, are possible and may be used to amplify from dissected organs; alternatively, sufficient mRNA may be extracted from later embryos for Northern blot analysis, although the small number of genes that can be investigated may prohibit this use.

cDNA libraries from stage-specific human embryos are being constructed in various centres (Jay *et al.*, 1997; Newcastle Human Embryos Gene Expression Group, unpublished data) and this also establishes a renewable source that may be distributed. Given the potential value of such libraries, it would seem appropriate to exploit micro-array technology to deposit these clones on cDNA chips in addition to using these advances to perform a whole-genome expression screen at each developmental stage.

There are clear advantages for combining these approaches with methods that

4.3 Cutting and storage

The precious nature of the tissue means that attempts are made to collect each cut section of an embryo and the aim is to detect one mRNA species or protein per section.

In the case of paraffin-embedded tissue, this may be achieved by microtome sectioning so that ribbons up to 40 cm in length are generated, which are laid onto suitable RNase free trays. After sectioning is completed, the paraffin sections can be mounted onto slides. This approach may be preferable to cutting short ribbons of four or five sections which would be floated onto slides immediately. The former approach keeps the duration of sectioning to a minimum and prevents a microtome 'bottleneck'. If an embryo is 'laid out' on a tray, individual sections can be removed for histological staining in order to identify anatomical structures to determine the number of sections to be placed per slide. This may be important if serial sections from one organ are to have alternate immunocytochemical and *in situ* hybridization detections whilst another embryonic region would be better mounted in ribbons of four sections. Apart from the precious nature of the human embryos, collecting all the sections would enable digital 3-D reconstruction if this were required. If computer 3-D reconstruction is planned, an issue to consider is registration of sections with respect to each other. One solution is to embed the embryo in a 'frame' of material so that registration points would surround the sectioned embryo on each slide to enable a positional reference from slide to slide. This has been a solution successfully employed by groups who have physically constructed models from microscopic sections. Although points of registration are also required by some current reconstruction computer programs, fortunately this is not always the case and section alignment can now be 'calculated' using more recent software (see Chapters 13 and 14).

4.4 Probe labelling

The labelling of a nucleotide for *in situ* hybridization may either use direct or indirect methods. Direct labelling incorporates a reporter molecule within the nucleotide, for instance ^{35}S or fluorescein, and indirect labelling encorporates a moiety within the nucleotide that may be localized by cytochemical techniques, for instance digoxigenin or biotin. For both methods the presence of the reporter should not significantly interfere with the synthesis or stability of the nucleotide nor affect hybridization to tissue mRNA. In most instances the nucleotide is labelled by enzymatic incorporation of a modified (d)NTP by polymerase 1, Taq DNA polymerase, terminal deoxynucleotidyl transferase, or a bacteriophage RNA polymerase (T7, T3, SP6).

A consideration when choosing the detection method is what the required levels of sensitivity and signal resolution are. It is generally considered that radioactive detection is more sensitive than non-radioactive methods and thus may be preferred if the gene investigated has a low level of expression. However, the signal from the more commonly used radioisotopes (such as ^{35}S or ^{33}P), does not provide the same resolution as that achieved using non-radioactive methods. This is demonstrated in *Figure 2*, which compares the expression of myosin light chain 2V (MLC-2V) in mouse embryos using ^{35}S- and digoxigenin-labelled RNA

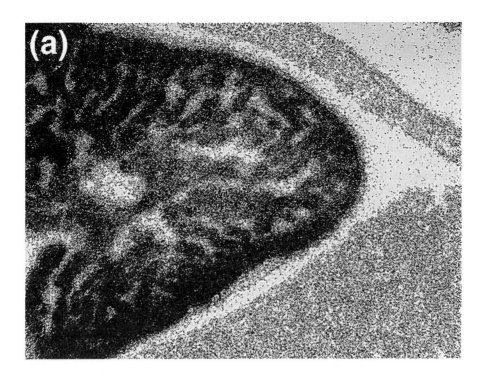

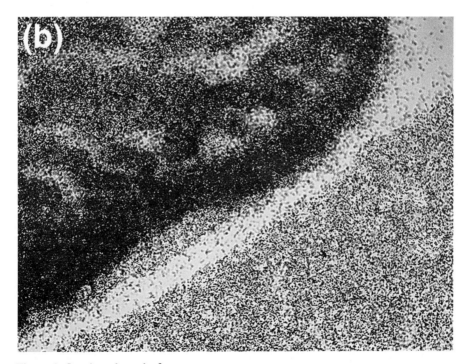

Figure 2. Continued overleaf.

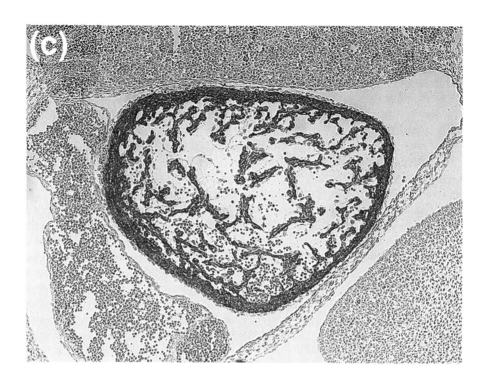

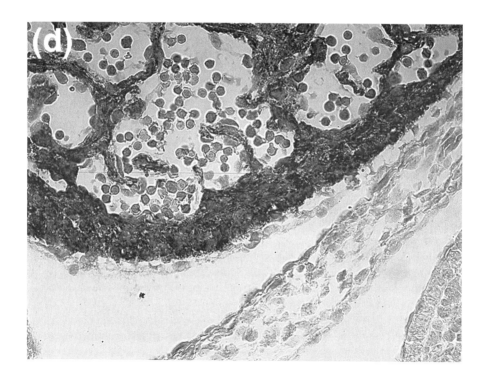

probes, respectively. Both give clear expression in the cardiac ventricle but the signal from the digoxigenin-labelled probe can be localized over individual cells. Using the digoxigenin-labelled probe it is possible to see that the cells overlying the ventricle (future pericardium) do not express MLC-2V, whereas this could not be determined with such clarity from the ^{35}S signal.

4.5 Pretreatments of tissue sections

The tissue sections are pretreated prior to hybridization, which may include dewaxing, HCl, detergent, protease, dithiothreitol, fixation and prehybridization steps. Protocols differ as to whether all steps are necessary and some seem to be included for empirical reasons rather than logical justification; for instance, the precise action of the HCl treatment step is unclear; it may denature or remove proteins from the section and so facilitate the binding of probe to target. However, as this step usually improves the signal/noise ratio, it is often included.

4.6 Hybridization, washes and detection

During hybridization, complementary nucleotide sequences anneal and this is primarily influenced by the hybridization temperature, monovalent cation concentration, organic solvents (formamide), probe length and GC content of the probe and its target. Hybrid formation is favoured by decreased temperature, high salt and low formamide concentrations, together with longer sequence with a high GC content. The conditions of hybridization and how it relates to Tm (temperature at which half the hybrids are annealed) is crucial in determining the levels of specific and non-specific binding (i.e. signal/noise). Formulae exist which can be used to estimate the optimal conditions but in practical terms the initial experimental conditions are empirical and determined by a laboratory's previous experience; this usually entails hybridizing and washing between 51 and 54°C in 50–60% formamide/1–2 x SSC.

The detection of hybridized nucleotide is obviously dependent on the labelling method and, as discussed previously is either by incorporation of radioisotopes or non-radioactive ligands. Radioisotopes (^{35}S or ^{33}P) may be detected either by radiographic film, which gives a rapid result but very poor resolution, or more usually, by photographic emulsion, which has improved resolution but may need up to a 14 day exposure time. If the signal is low, a longer exposure time may be required or, in the case of tritium, which gives better tissue resolution, exposure times of 3

Figure 2. Tissue *in situ* hybridization which compares the resolution of radioactive and non-radioactive probe labelling techniques. Mid-saggital sections of a 12 embryonic day mouse embryo showing the cardiac ventricles hybridized with antisense riboprobes for myosin light chain 2V (MLC-2V). (a) and (b) (higher magnification): MLC-2V probe labelled with ^{35}S; (a) shows strong signal over the ventricle. (c) and (d) (higher magnification): MLC2V probe labelled with digoxigenin. Strong expression is present in the ventricles and can be localized to the trabeculae. It is clear that the single cell layer covering the ventricle (future pericardium) does not express MLC-2V; this is not so evident from the ^{35}S-labelled probe.

months are not uncommon. Non-radioactive methods of detection are promoted for reasons of safety, convenience, expense, speed with which results are generated, and the cellular resolution of signal (*Figure 2*).

4.7 Cycling tissue in situ

Recently a new method has been devised, which localizes and detects mRNA within tissue sections, combining RT–PCR with tissue *in situ* hybridization (Martinez *et al.*, 1995; Mee *et al.*, 1996a,b). It is claimed that this increases sensitivity and so may detect rare mRNA species which are otherwise below the level of detection using standard tissue *in situ* hybridization. Following tissue fixation, sectioning and hybridization pretreatments (e.g. dewaxing, proteinase permeabilization) using standard techniques, reverse transcription is performed by incubating the tissue sections on glass microscope slides with RT and random hexanucleotides to generate first strand cDNA. A PCR mixture is then applied to the section, including primers specific to the gene of interest and labelled nucleotides (e.g. digoxigenin-dUTP), then slide thermal cycling performed. Various machines are available commercially for this purpose, but the number of slides that can be processed simultaneously may be limited. Unbound reagents are then washed off and incorporated digoxigenin detected using standard methods. Additional negative controls have to be included; for instance, no RT, no primers and no Taq-1 polymerase.

This method has the potential advantage of the increased sensitivity that PCR amplification offers and there have been reports indicating that it is more sensitive than standard radioactive tissue *in situ* protocols (Mee *et al.*, 1996a). Another advantage is that separate members of a gene family may be amplified if primers are chosen which are specific to the individual gene; cDNA or riboprobes may have homologous regions that will reanneal to mRNA of other genes; for instance, if they contain homeobox or pairbox sequences. Whilst it is usual to synthesize a riboprobe from a template that does not have a common motif, it may be difficult to be certain of the specificity of the probe. Primer specificity may be confirmed by PCR amplification from cDNA extracted from tissues, and confirming that there is only one product.

There are, however, potential disadvantages, apart from the expense of the thermal slide cycler and the limited number of slides that may be processed in one experiment. There are many artefacts that have been reported with standard PCR and it is possible that amplification of sequences within tissue sections by cycling *in situ* also generates artefacts; strict controls have, therefore to be included with each reaction in addition to those of standard tissue *in situ* hybridization. These necessary controls will reduce the number of genes that may be investigated per embryo.

A further consideration is whether the technique is too sensitive and may detect 'surrogate' transcription. It is evident that genes may not be switched off completely, so that mRNA may be generated from 'leaky' transcription of all genes in all tissues. Genes have been amplified from mRNA extracted from peripheral blood lymphocytes which would not be not considered to have 'functional mRNA expression' in such cells (Roberts *et al.*, 1991). If cycling *in situ* were able to amplify from surrogate transcripts, interpretation of results would, at best, be very difficult and, at worst, meaningless.

4.8 Data and resource storage

An issue that rapidly becomes apparent when establishing a human embryo resource is the logistical problem of storage. The storage of the embryos is obviously crucial and if embedded in paraffin blocks they may be kept at 4°C. However, once sectioned they occupy a much more significant amount of space and boxes filled with microscope slides fill 4°C fridges or cold rooms. A Carnegie Stage 12 embryo transversely cut, four sections per slide, could require 100 slides (one box), whereas a Carnegie Stage 23 embryo similarly cut and mounted would require 1500 slides (15 boxes). As there is justification for collecting each section cut, a laboratory could rapidly fill with slide boxes. The logistics of using frozen embryos for cryosectioning is perhaps even worse. Paraffin-embedded tissue is usually kept at 4°C if the sections are to be used for RNA *in situ*. However, paraffin is an inert medium which means that, if necessary, sections could be stored at room temperature. This is not the case for cryosectioned material. Once cut, the sections have to kept frozen, preferably at −80°C. The space and costs of −80°C freezers, with CO_2 backup, to store any significant number of embryos is large.

The storage of data is also an important consideration. The data relating to the staging, handling, fixation, cutting and use of the embryos needs to be carefully documented. The information stored needs to conform with any restrictions regarding confidentiality; in the UK data that may identify the woman who agreed to the embryo being used is not held by the researchers. The details of each experiment and the results obtained need to be recorded and kept in appropriate relational databases in order that they can be retrieved.

5. Conclusion

The necessity of studying human embryonic material is being recognized as important for the understanding of human development. Whilst there are similarities between the development of human and laboratory species, the differences that exist dictate that humans embryos are required for comparison in order to identify the distinctions between humans and other mammalian species. The methods that can be used in human embryos are fundamentally no different from those used in other species; however, methodological differences or restrictions are imposed due to the availability and use of human tissue. The validity of the results from any group working in this field requires independent confirmation; it is hoped that interest in gene expression in human development is sufficient to stimulate more groups to use this approach and so results may be compiled and compared. This will be facilitated by making use of increasingly powerful database and image analysis tools available.

References

Adams MD. (1996) Serial analysis of gene expression: ESTs get smaller. *Bioessays* **18**: 261–262.
Akhurst RJ. (1992) Localisation of growth factor mRNA in tissue sections by *in situ* hybridization. In: *Growth Factors. A Practical Approach* (eds I McKay, I Leigh). IRL: Oxford University Press, Oxford, pp. 109–132.

8

Biological studies of preimplantation human embryos

Virginia Bolton

1. Introduction

1.1 Historical perspective

The earliest studies of preimplantation embryos relied upon the limited material obtained from hysterectomy specimens (Hertig *et al.*, 1954). In the two decades since the development of techniques for the fertilization of human oocytes *in vitro*, (Edwards *et al.*, 1969, 1970), the increased accessibility of the human embryo for study has led to a rapid increase in understanding of the earliest stages of human embryogenesis. However, ethical and legal constraints on the nature of investigation (see below and also Chapter 4), together with the fact that human embryos are generated *in vitro* almost exclusively for use in therapeutic *in vitro* fertilization (IVF), rather than for research, mean that the number of embryos available for research remains limited. Consequently, the rate of advance in our knowledge of early development in the human remains relatively slow, compared with that in other mammalian embryos.

Although the laws governing research using human preimplantation embryos vary worldwide, it is generally true that it is either prohibited, or strictly regulated. Currently in the UK, IVF is regulated by the Human Fertilisation and Embryology Authority, a statutory body established following the passing of the *Human Fertilisation and Embryology Act (1990)*. Research using preimplantation human embryos is permitted only under licence, up to a maximum of 14 days after fertilization, and only provided the studies aim to achieve one or more of the following:

(i)   To promote advances in the treatment of infertility
(ii) To increase knowledge about the causes of congenital disease
(iii) To increase knowledge about the causes of miscarriages
(iv) To develop more effective techniques of contraception
(v)   To develop methods for detecting the presence of gene or chromosome abnormalities in embryos before implantation.

Molecular Genetics of Early Human Development, edited by T. Strachan, S. Lindsay and D.I. Wilson.
© 1997 BIOS Scientific Publishers Ltd, Oxford.

To date, the majority of studies utilizing supernumerary human embryos donated to research have aimed to improve the success rate of therapeutic IVF, through improved understanding both of the development of embryos *in vitro* and of mechanisms for selecting viable embryos for transfer in order to achieve a successful pregnancy. Other studies, including those involving techniques for preimplantation diagnosis of genetic disease, will not be considered here (see Chapter 9).

1.2 Development of the preimplantation human embryo

Following insemination *in vitro*, the spermatozoon penetrates the oocyte, meiosis II is resumed and the second polar body is extruded. The timing of specific events during the first cell cycle of human embryogenesis has been examined in some detail (summarized in *Figure 1*; Capmany *et al.*, 1996), and it is now known that pronuclei develop between 3 and 10 h post-insemination (median 8 h), and that DNA replication occurs between 8 and 22 h post-insemination. Although male and female pronuclei are morphologically similar (Wiker *et al.*, 1990), they form at different sites, with the male pronucleus normally forming near the point of sperm entry, and the female pronucleus forming at the ooplasmic pole of the meiotic spindle. From these positions they migrate towards the centre of the oocyte, enlarge and flatten progressively against each other prior to fusion (Wright *et al.*, 1990).

The duration of M-phase is relatively constant and lasts 3–4 h, whereas the times of pronuclear breakdown and first cleavage division vary considerably between embryos. However, the majority of embryos undergo pronuclear breakdown and first cleavage between 27 and 30 h, and between 29 and 32 h post-insemination respectively (Capmany *et al.*, 1996). Subsequent cleavage divisions, to the

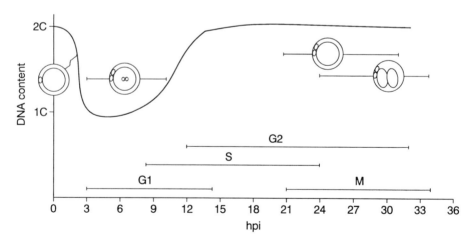

Figure 1. Diagram summarizing the timing of pronuclear formation and breakdown, DNA synthesis and cleavage during the first cell cycle of human embryogenesis. hpi, h post-insemination; 1C, haploid DNA content; 2C, diploid DNA content. Redrawn from Capmany *et al.* (1996) by permission of Oxford University Press.

4-, 8- and 16-cell stages, which are completed within 45, 69 and 93 h of insemination respectively (Braude *et al.*, 1990) have not been investigated in such detail. Similarly, there is little information concerning the generation of polarity in the human embryo, a process which is fundamental to the establishment of the first two cell lineages, the inner cell mass (ICM) and trophectoderm (TE) in the preimplantation mammalian embryo (see reviews by Johnson and Maro, 1986; Maro *et al.*, 1991). Scanning electron microscopy (SEM) has shown that in contrast to other mammalian species studied, the human oocyte has no microvillus-free pole overlying the metaphase spindle, and the plasma membrane appears to be functionally homologous with respect to sperm entry sites (Santella *et al.*, 1992). Moreover, following fertilization, the human oolemma does not appear to undergo any major reorganization in terms of redistribution of microvilli, at least up to the 8-cell stage (Dale *et al.*, 1991, 1995), which again contrasts with what is known to occur in the mouse (Johnson *et al.*, 1975). Lineage analysis has shown that commitment of cells to one of the two cell types of the blastocyst does not occur prior to the 8-cell stage (Mottla *et al.*, 1995). Despite these differences, it seems unlikely that the mechanism by which cell diversity is generated in the preimplantation human embryo differs greatly from that of other mammalian species, although the timing may differ.

Relatively few studies have examined the human embryo directly during the transition from morula to blastocyst, which occurs between 4 and 6 days following fertilization *in vitro*. However, animal studies have shown that this stage of development is characterized by major differentiative events that differ between the ICM and TE cells, and it is likely that these findings can be extrapolated to the human. In the mouse, the two cell lineages show increasing divergence in morphology, including cell shape and membrane specializations (Ducibella *et al.*, 1975; Nadijcka and Hillman, 1974), in transport mechanisms (Fleming and George, 1987; see review by Borland, 1977), and intercellular junctions, allowing the formation and retention of the blastocoelic cavity (Magnuson *et al.*, 1977), and in tissue-specific patterns of polypeptide synthesis (Handyside and Johnson, 1978). Gap junctions are first seen at the 4-cell stage in the human embryo (Hardy *et al.*, 1996), while in blastocysts, desmosomes and gap junctions have been identified between ICM cells, and between ICM and TE cells, and apical tight junctions have been identified between TE cells (Sathananthan *et al.*, 1990).

1.3 Survival of human embryos in vitro

Reports of the number of human embryos derived by IVF that continue development to the blastocyst stage *in vitro* range from 0 to 82% (see *Table 1*). This wide range of findings reflects not only the different media used for culture, but also the different times at which embryos were examined, the use of subjective morphological evaluation of blastocysts, as well as the selective inclusion of early, expanding and fully expanded blastocysts. Despite these limitations, there is no doubt that a relatively small number of human embryos survive to the fully expanded blastocyst stage *in vitro*. Although the average of 34% from these studies compares favourably with the 29% (11/38) of embryos collected after *in vivo* fertilization and uterine lavage (Buster *et al.*, 1985; Formigli *et al.*, 1987), the different

Table 1. Summary of results from studies in which human embryos were cultured to the blastocyst stage

Authors	Embryos cultured (n)	Blastocysts formed [n(%)]	Culture medium	Description of embryos scored as blastocysts
Dawson et al., 1988	37	21 (57)	T6+10% patient serum	Blastocyst
Bolton et al., 1989	315	55 (17)	EBSS+10% patient serum	Fully expanded
Bongso et al., 1989	18	5 (28)	T6+15% human serum	Fully expanded
Hardy et al., 1989b	149	63 (42)	T6 or EBSS+10% patient serum	Expanded
Hardy et al., 1989a	43	10 (23)	T6+10% patient serum	Blastocyst
Lopata and Hay, 1989	392	104 (27)	MEM+10% human cord serum	Expanded
	234	34 (15)	HTF+1% human cord serum	Blastocyst
Muggleton-Harris et al., 1990	33	27 (82)	Ham's F12+10% FBS	Morula or expanded blastocyst
	5	1 (20)	EBSS+10% FBS	
	3	0 (0)	Eagle's+10% FBS	
	3	0 (0)	M16+10% BSA	
	6	1 (17)	BRL-CM+10% FBS	
Bolton et al., 1991	197	78 (40)	EBSS+10% patient serum	Cavitating, expanding and fully expanded blastocyst
Dokras et al., 1991	172	47 (27)	T6+10% patient serum	Expanded
Winston et al., 1991	159	62 (39)	EBSS+10% patient serum	'Blastocyst-like'
Fitzgerald and DiMattina, 1992	32	18 (56)	CZB+10% patient serum	Blastocyst
	24	5 (21)	EBSS+10% patient serum	Blastocyst
Dokras et al., 1993	573	151 (26)	T6+10% patient serum	Expanded
Van Blerkom, 1993	127	49 (39)	EBSS+15% patient serum	Expanded

T6, Tyrode's 6 medium; EBSS, Earle's balanced salt solution; MEM, minimal essential medium; HTF, human tubal fluid (medium); FBS, fetal bovine serum, BSA, bovine serum albumin; BRL-CH, buffalo rat liver-conditioned MEM; CZB, Chatot, Ziomek and Bavister medium.

survival rates in different media, together with the range of success rates achieved by different centres offering therapeutic IVF (range: 0–23.8% per treatment cycle; HFEA, 1996), mean that factors both extrinsic and intrinsic to the embryo influence its survival. Thus, in terms of improving the success rate of therapeutic IVF, research using human embryos has examined why the human embryo so frequently undergoes arrest prior to blastocyst formation. Attempts have been made to develop both conditions that will optimally support human embryogenesis *in vitro*, and techniques by which embryos that are most likely to continue development can be identified.

2. Culture conditions

2.1 Culture media

The culture media used for human IVF have been adapted from those developed originally for culture of animal embryos, primarily the mouse and rabbit (see review by Biggers, 1987). Although these media can clearly support human embryogenesis, it is likely that they provide a suboptimal environment. Evidence cited to support this includes the high incidence of developmental arrest (Bolton *et al.*, 1989), uneven cleavage, cytoplasmic fragmentation, and anucleate or multinucleate blastomeres (Winston *et al.*, 1991), yet studies of *in vivo*-derived human embryos collected after uterine lavage for the purpose of oocyte donation have shown similar developmental anomalies (Buster *et al.*, 1985; Formigli *et al.*, 1987). Thus, while it is clear that human embryos vary inherently in their potential, their inherent viability may be further compromised by sub-optimal conditions *in vitro*.

It is known that the potential of cleavage stage human embryos to continue development is reflected, at least in part, by their gross morphology, with those embryos that display regular, spherical blastomeres and no fragmentation surviving to the fully expanded blastocyst stage more frequently than those with irregular blastomeres and fragmentation (Bolton *et al.*, 1989; Dokras *et al.*, 1993). The rate of cleavage may also be an indicator of embryo viability, with those embryos that cleave to the 4-cell stage by day 2 following insemination *in vitro* forming blastocysts on day 5 more frequently than embryos that develop either more slowly or, interestingly, more quickly (Bolton *et al.*, 1989; Dokras *et al.*, 1993). Thus, while a number of studies have attempted to evaluate different culture systems, or develop defined culture media that will support human embryogenesis to the blastocyst stage and beyond, this heterogeneity among populations of human embryos means that interpretation of results of such studies is difficult, unless embryo morphology is described in detail. Moreover, although many blastocysts which develop *in vitro* are abnormal (Dokras *et al.*, 1993; Hardy *et al.*, 1989a), as reflected in the failure to establish pregnancy after transfer to the uterus (Bolton *et al.*, 1991), few studies reporting rates of blastocyst formation validate the normality of blastocysts that form (*Table 1*).

Using the simple media, Tyrode's 6 (T6) and Earle's balanced salt solution (EBSS), the reported rate of blastocyst formation varies considerably (17%: Bolton *et al.*, 1989; 26%: Dokras *et al.*, 1993; 42%: Hardy *et al.*, 1989a). However, since up to 24% of blastocysts that develop are abnormal, in terms of morphology, nuclear number, cell allocation to ICM and TE, and levels of secretion of human chorionic gonadotrophin (hCG; Dokras et al., 1993; Hardy et al., 1989a), it has been claimed that simple media are unsatisfactory for human IVF and embryo culture (Lopata, 1992).

Complex media contain factors additional to the basic salts of simple media, such as hypoxanthine, glutamine, vitamins and essential amino acids (Ham, 1965). One study reported that 85% of embryos formed morulae or blastocysts in the complex medium Ham's F12; although the validity of this finding is thrown into question by the small sample size (Muggleton-Harris et al., 1990). More extensive studies have been carried out using the complex medium, alpha-mini-mal essential medium (α-MEM) supplemented with 1% human cord serum (HCS). This supports a blastocyst formation rate of up to 60% (Lopata, 1992), with blastocysts secreting hCG at levels 20-times greater than those grown in simple media, and increasing daily up to day 14 after insemination (Lopata and Oliva, 1993a,b), rather than decreasing beyond day 10 (Dokras et al., 1991, 1993).

Human tubal fluid medium (HTF; Quinn et al., 1985), a further simple medium that is used for human IVF and embryo culture, is based on the composition of human oviductal fluid (Borland et al., 1980) and contains more potassium, more calcium and less glucose than T6 (Table 2). Despite the findings of a randomized clinical trial using either HTF or T6 supplemented with 10% patient serum for therapeutic IVF, where more pregnancies were achieved using HTF than T6 (18/60 compared with 6/53; Quinn et al., 1985), the findings of culture studies were less promising: fewer blastocysts developed when spare human embryos were cultured in HTF containing 1% HCS compared with α-MEM with or without serum (14.5% compared with 26.5%), and the blastocysts synthesized lower levels of hCG (Lopata and Hay, 1989). Moreover, the clinical findings were

Table 2. Composition of Earle's balanced salt solution (EBSS), human tubal fluid (HTF), Tyrode's 6 (T6) and complex serum-free medium (CSFM)

Component (mM)	EBSS	HTF	T6	CSFM
NaCl	117.0	101.6	99.4	110
KCl	5.3	4.69	1.42	4.78
$MgSO_4.7H_2O$	0.8	0.2	0.71	–
$Mg\ Cl_2.6H_2O$	–	–	–	1.0
KH_2PO_4	–	0.37	–	–
NaH_2PO_4	1.0	–	0.36	1.2
$CaCl_2.2H_2O$	1.8	2.04	1.78	1.8
$NaHCO_3$	26.0	25.0	25.0	25.0
$Na_2SO_4.10H_2O$	–	–	–	1.0
Glucose	5.6	2.78	5.56	3.4
Na^+pyruvate	0.8	0.33	0.47	0.37
Na^+lactate	–	21.4	24.9	5.0
Glutamine	–	–	–	1.0
Non-essential amino acids[a]	–	–	–	10 ml l^{-1}
Vitamins[b]	–	–	–	10 ml l^{-1}
Human apo-transferrin	–	–	–	25 mg ml^{-1}

[a,b] 100 x MEM Eagle's, commercial preparation (Eagle, 1959).

not borne out by a subsequent trial, which found no differences between the two media (Cummins *et al.*, 1986a).

The use of serum introduces an undefined variable into studies of embryo culture; for example, while the composition of HTF aims specifically to reduce the level of glucose to which the human embryo is exposed *in vitro*, glucose present in patient serum may reverse this effect. Indeed, serum contains a large, undefined complement of factors, some of which may have inhibitory and others beneficial effects, and when patient serum is used this variable is different for each cohort of embryos studied. This concern has led to the use of synthetic serum substitutes (Holst *et al.*, 1990), or commercially prepared serum products in therapeutic IVF, but their efficacy has been questioned (Ashwood-Smith *et al.*, 1989; Lopata, 1992).

While any investigation of the effect of different components of culture media on development of human embryos requires the use of a serum-free medium, omitting serum from the culture system impairs embryo development; for example, when human embryos were cultured for 2 days in the complex Menezo's B2 medium without serum (Menezo *et al.*, 1984), the damage they sustained was not repaired after return to medium containing 15% human serum (Menezo *et al.*, 1990). In an attempt to overcome this problem, a complex, serum-free medium (CSFM) has been described, the constituents of which are based on a range of studies using animal and human embryos (Dunglison *et al.*, 1996; see *Table 2*). The development of human blastocysts in CSFM demonstrates that serum is not essential for human embryogenesis *in vitro* (Dunglison *et al.*, 1996) but neither blastocyst formation rate, nor the viability of the blastocysts which formed was as good as in T6 medium supplemented with 10% patient serum (Dokras *et al.*, 1991, 1993). Nonetheless, the use of CSFM will enable investigation of the effects of individual growth-promoting and protecting factors which animal studies suggest may have roles in human embryogenesis. Such factors may include insulin (Harvey and Kaye, 1990), insulin-like growth factors (IGFs; Gardner and Kaye, 1991), epidermal growth factor (EGF; Wood and Kaye, 1989), colony-stimulating factor-1 CSF-1; Paria and Dey, 1990), platelet-derived growth factor (PDGF; Larson *et al.*, 1992), thioredoxin and superoxide dismutase (Nonogaki *et al.*, 1991), ethylenediaminetetraacetic acid (EDTA; Abramczuk *et al.*, 1977) and leukaemia-inhibitory factor (LIF; Fry *et al*, 1992; Marquant-Le Guinne *et al.*, 1993; Robertson *et al.*, 1994; summarized by Dunglison *et al.*, 1996; Lopata, 1992).

2.2 Group culture and micro-drop culture

It is known that the mammalian embryo secretes certain factors which, it is thought, may have autocrine effects on its development. In the human, factors secreted by the preimplantation embryo include transforming growth factor-α (TGF-α) and IGF-II (Hemmings *et al.*, 1992), PDGF (Svalander *et al.*, 1991), and platelet-activating factor (PAF; Collier *et al.*, 1990; Nakatsuka *et al.*, 1992). In the mouse, evidence to support the autocrine effect of embryo secretions comes from studies where embryos were cultured in microdrops or in groups, in order to concentrate, rather than dilute, any factors that may be secreted (Cansecoe *et al.*, 1992; Lane and Gardner, 1992; Paria and Dey, 1990).

A preliminary *in vitro* study in the human failed to replicate these findings (Dunglison *et al.*, 1996), although two small clinical studies found that group culture enhanced both cleavage rate and embryo morphology (Moessner and Dodson, 1995), and consequently pregnancy rate following embryo transfer (Almagor *et al.*, 1996).

2.3 Co-culture with feeder cells

There have been many studies which report improved implantation rates after transfer of human blastocysts that develop during co-culture with feeder cells. These include those using Vero cells derived from African green monkey kidneys (Menezo *et al.*, 1992; Olivennes *et al.*, 1994; Plachot *et al.*, 1995), fetal bovine fibroblasts (Wiemer *et al.*, 1989a,b), bovine oviductal epithelial cells (Wiemer *et al.*, 1993), human ampullary cells (Bongso *et al.*, 1989), homologous endometrial cells (Birkenfeld and Navot, 1991), homologous cumulus cells (Mansour *et al.*, 1994), granulosa cells (Freeman *et al.*, 1995) and ovarian cancer cells (Ben-Chetrit *et al.*, 1996; see review by Thibodeaux and Godke, 1995). Despite these reports, the use of co-culture for therapeutic IVF remains controversial, and while there are co-culture studies which report blastocyst formation rates of over 60% (Bongso *et al.*, 1989; Menezo *et al.*, 1992) and correspondingly increased implantation rates following embryo transfer (Olivennes *et al.*, 1994), others report no significant difference between embryos co-cultured with feeder cells and those cultured alone (Sakkas *et al.*, 1994; Van Blerkom, 1993). Indeed, among the multitude of studies reported in the literature, very few have been undertaken prospectively, with other variables controlled, and only these enable an objective, scientific evaluation of the effects of co-culture on preimplantation human embryo development (Bavister, 1992). Three such studies suggest that not only do significantly more embryos develop into fully expanded blastocysts (Ben-Chetrit *et al.*, 1996; Quinn and Margalit, 1996) but also a greater proportion of blastocysts commence hCG secretion (Turner and Lenton, 1996) when cultured with feeder cells. It has also been reported that the presence of feeder cells may assist some embryos in recovery from delayed cleavage (Vlad *et al.*, 1996). Such beneficial effects of feeder cells are presumably due to the secretion of growth-promoting factors and/or removal of embryo-toxic factors (see review by Ryan and O'Neill, 1994).

Despite its possible advantages, the technique is expensive and time-consuming, and does not contribute to a greater understanding of the *in vitro* requirements of the human embryo (Bavister, 1992). Moreover, it is not without risk, as it is possible that contaminated cultures may result in the transfer of disease (Olivennes *et al.*, 1994). Thus, while co-culture may provide short-term advantages, in terms of success rates of therapeutic IVF, it remains preferable, in terms of an overall increase in understanding of human embryogenesis, to attempt to produce defined media for human embryo culture.

3. Assessment of human embryo viability

3.1 Morphology

Morphological criteria, currently used in the selection of embryos for transfer during therapeutic IVF, are, at best, limited (Bolton et al., 1989). Moreover, the practice of transferring multiple embryos, up to a maximum of three in the UK, represents a major obstacle in defining valid tests of embryo viability. There have been a number of attempts to derive embryo scoring systems, based on cleavage rate, symmetry of blastomeres, appearance of the cytoplasm and degree of extra-cellular fragmentation, to predict embryo viability (Cohen et al., 1988; Cummins et al., 1986b; Shulman et al., 1993). It has been reported that cytoplasmic frag-mentation appears to be the most important variable in predicting embryo viabil-ity, followed by cleavage rate, blastomere symmetry and cytoplasmic granularity (see review by Clarke, 1995).

Analysis of IVF cycles in which a single embryo was transferred yields the most accurate information concerning the relationship between embryo morphology and viability. A series of 957 such cycles, where a single embryo was transferred either because only one was obtained ($n = 931$) or at the request of the patient ($n = 26$), confirmed the association between blastomere symmetry and lack of cytoplasmic fragmentation and viability. In addition, it demonstrated that embryos which had cleaved to the 4-cell stage by 48 h following insemination implanted more frequently than those which cleaved either more slowly or more rapidly (Giorgetti et al., 1995). This supports suggestions from earlier studies that there appears to be an optimum rate of cleavage for human embryos (Bolton et al., 1989; Cummins et al., 1986b; Testart, 1986).

While scoring systems based on morphology enable the establishment of statis-tical trends among populations of embryos, they do not give an accurate predic-tion as to the viability of individual embryos. The failure of any such embryo scoring system yet developed to provide an accurate prediction of implantation highlights the need for additional criteria for evaluating embryo quality.

3.2 Embryo metabolism

During the earliest stages of preimplantation mammalian embryogenesis, energy is generated by the oxidation of substrates such as pyruvate, lactate, amino acids and possibly fat, while blastocyst formation requires a sharp increase in glucose con-sumption (see review by Leese, 1995). Sensitive techniques for the measurement of nutrient uptake or efflux using ultramicrofluorimetry (Mroz and Lechene, 1980), adapted for use with preimplantation mammalian embryos (Leese and Barton, 1984), have enabled the measurement of metabolic activity of human embryos in vitro (Hardy et al., 1989b; Leese et al., 1986). As non-invasive assays, it was hoped that they might allow the objective assessment of embryo viability prior to transfer in therapeutic IVF. Although they have proved to be of little use in predicting embryo potential up to day 3 after insemination (Conaghan et al., 1993), studies of glucose and pyruvate uptake, and of lactate production have demonstrated significant differ-ences on days 5 and 6 between arrested embryos and those that form blastocysts (Gott et al., 1990). It has been suggested that, while measurement of pyruvate uptake alone up to day 3 after insemination is unlikely to provide a useful criterion by which to assess embryo viability, measurement of all three parameters on day 4 after insem-ination, by which time differences between viable and non-viable embryos appear to be more pronounced, may be more informative (Conaghan et al., 1993).

3.3 Embryo secretions

The suggestion that the cleavage stage mammalian embryo secretes factors that may have autocrine or paracrine effects (Paria and Dey, 1990), and the certainty that the peri-implantation embryo must communicate with its maternal host, open the possibility of identifying specific factors, and measuring their levels of secretion, as alternative non-invasive assays of embryo viability. The level of hCG secretion by individual human blastocysts has been used as a means of assessing their potential (Dokras et al., 1991; Lopata and Oliva, 1993a,b; Turner and Lenton, 1996); its production in vitro is time-dependent (Woodward et al., 1993) and levels do not necessarily correlate with the morphology or cleavage rate of the embryo from which the blastocyst develops (Woodward et al., 1994). The earliest detectable hCG secretion by human blastocysts has been reported to be the evening of day 7 after insemination, using an immunoradiometric assay (Lopata and Hay, 1989), 160 h post-insemination, using a microparticle enzyme immunoassay (Woodward et al., 1993), and 6 days following insemination, using a mouse Leydig cell bioassay (Jones et al., 1992). Thus, while the use of assays for hCG secretion for the selection of embryos for transfer during therapeutic IVF seems impractical, it is undoubtedly of value in the assessment of the viability of blastocysts which develop under different culture conditions (Lopata and Oliva, 1993b; Turner and Lenton, 1996).

A further hormone which may prove predictive of embryo potential is pregnancy specific β_1-glycoprotein (SP-1), which is known to be synthesized by the syncytiotrophoblast (Horne et al., 1976) and has been detected in culture medium from preimplantation human embryos as early as day 1 after insemination (Dimitriadou et al., 1992; Saith et al., 1996). A preliminary study which measured SP-1 secretion by human blastocysts prior to transfer during therapeutic IVF suggested that those embryos that implanted secreted higher levels of the protein than those that did not (Bolton, 1993). However, this finding was not substantiated by a larger in vitro study, which found an inconsistent pattern of SP-1 secretion by human blastocysts of varying quality, assessed morphologically (Saith et al., 1996).

A number of different assay systems have been used to investigate the secretion of growth factors by human embryos in vitro. Bioassays have demonstrated the secretion of CSF-1, interleukins 1 and 6 (IL-1, IL-6), tumour necrosis factor-α (TNF-α) and TGF-β by human embryos between the 2-cell and 8-cell stages, although the different studies vary in the levels of secretion reported (Austgulen et al., 1994; Zolti et al., 1991). Enzyme-linked immunosorbent assays (ELISA) failed to detect soluble IL-6 receptor in human embryo culture supernatants (Austgulen et al., 1994), but confirmed the secretion of IL-1 (Sheth et al., 1991) and TNF-α (Witkin et al., 1991). Radioimmuno- and radioreceptor assays have detected secretion of TGF-α and IGF-II by morula and blastocyst stage embryos (Hemmings et al., 1992).

Among the cytokines, there are a number of potential candidates for a putative embryonic signal to the endometrium or corpus luteum. Interferon-α might be expected to play a role in signalling between the embryo and corpus luteum, from studies of ruminants (see review by Chard, 1991); however, it has not been detected in two studies which measured levels in medium in which human blastocysts developed (Bolton, 1993; Gunn et al., 1994). Other possible factors include PDGF, α-immunoreactive inhibin, immunosuppressive factors and PAF. Preliminary reports of PDGF (Svalander *et al.*, 1991) and α-immunoreactive inhibin (Phocas *et al.*, 1992) secretion by human blastocysts require further investigation. Similarly, although there have been reports of immunosuppressive activity in medium used for culture of human embryos (Daya and Clark, 1986a,b), with an association between absence of activity and implantation failure (Clark *et al.*, 1989), other studies have failed to confirm this association (Armstrong *et al.*, 1989; Segars *et al.*, 1989).

In contrast, PAF has been investigated more extensively, and its significance remains the subject of some controversy. PAF is a soluble factor secreted by the preimplantation mammalian embryo, and has been shown to be responsible for thrombocytopenia during early pregnancy in the mouse (O'Neill, 1985). It has been shown to be secreted by the human embryo *in vitro* (O'Neill *et al.*, 1987) as early as the 1-cell stage (Collier *et al.* 1988), and, while it may have a role in human embryogenesis before implantation and may serve as a useful marker for fertilization and the developmental stage of the embryo (Nakatsuka *et al.*, 1992), the predictive nature of PAF secretion in terms of embryo viability and potential for implantation, remains unclear. This is largely because of the limitations of the bioassay used for its measurement, which is difficult to validate (see review by Clarke, 1995). Any further evaluation of the role of PAF, and of its predictive value in the selection of embryos for transfer during therapeutic IVF, will rest on the development of a specific, quantitative assay.

4. Cytogenetic studies

4.1 Chromosomal abnormalities and fertilization

There have been many studies attempting to investigate the association between chromosomal abnormalities in the human oocyte and the failure to achieve fertilization *in vitro*. However, few have distinguished between sperm-related and oocyte-related factors in fertilization failure, leading to reports of abnormality rates among oocytes that fail to fertilize *in vitro* ranging from 15.4 to 58.7% (*Table 3*). In the one study which examined exclusively oocytes that failed to develop pronuclei due to oocyte-related factors (Almeida and Bolton, 1993), it was found that 29.5% of 237 such oocytes were immature; 9.3% had in fact been penetrated by spermatozoa which, due to immaturity of the oocyte's cytoplasm, had undergone premature chromosome condensation (PCC) and pronuclei had not formed. The remaining immature oocytes displayed nuclear immaturity, with the chromosomes arrested in the decondensed state. Of the 167 mature oocytes that failed to become fertilized, 58.7% were chromosomally abnormal, which is a figure considerably higher than earlier reports (*Table 3*). While it is possible that some of the variation between studies may be due to the use of different ovarian stimulation regimens prior to IVF, there appears to be no consistent relationship between stimulation regimen and either the incidence of immature oocytes (Ma *et al.*, 1989; Michaeli *et al.*, 1990; Pieters *et al.*, 1989; Schmiady and Kentenich, 1989) or the incidence of chromosomal abnormality in mature oocytes (Bongso *et al.*, 1988; Plachot *et al.*, 1988).

Table 3. Summary of results from cytogenetic studies in oocytes that failed to develop pronuclei after insemination *in vitro*

	Oocytes [n (%)]		Analysed oocytes [n (%)]	
Authors	Examined	Analysed	Immature	Abnormal chromosomes
Martin et al. (1986)	50	50 (100)	–	17 (34)
Veiga et al. (1987)	216	117 (54)	–	31 (26.5)
Bongso et al. (1988)	302	251 (83)	–	59 (23.5)
Djalali et al. (1988)	150	96 (64)	–	37 (38.5)
Pellestor and Sele 1988)	371	188 (50.7)	–	35 (18.6)
Plachot et al. (1988)	316	316 (100)	–	82 (26)
Ma et al. (1989)	94	65 (69)	11 (16.9)	31 (47.7)
Pieters et al. (1989)	285	104 (36.5)	1 (0.009)	16 (15.4)
Tarin et al. (1991)	262	168 (64)	–	32 (19)
Edirisinghe et al. (1992)	193	103 (53.4)	17 (16.5)	26 (25.2)
Almeida and Bolton (1993)	293	237 (80.9)	70 (29.5)	98 (58.7)

A comparison of the incidence of abnormality among oocytes that could be considered 'unfertilizable' with that among an unselected population of human oocytes has enabled an examination of the relationship between chromosomal abnormality in the oocyte and its ability to become fertilized. In a study of 215 oocytes that failed to fertilize following insemination with dysfunctional spermatozoa, 47% were found to be chromosomally abnormal (Almeida and Bolton, 1994). From these data, it has been calculated that, following ovarian stimulation, 26.6% of human oocytes that have the capacity to develop pronuclei following insemination *in vitro* will be chromosomally abnormal, compared with 20.4% of oocytes that are 'unfertilizable'. Statistical analysis demonstrates no significant difference between these figures, suggesting that there is no association between chromosomal abnormality in the oocyte and its ability to become fertilized *in vitro* (Almeida and Bolton, 1994).

4.2 Chromosomal abnormalities in cleavage-stage embryos

The calculated incidence of 26.6% chromosomal abnormality among oocytes that have the capacity to develop pronuclei following insemination (Almeida and Bolton, 1994), together with the estimated chromosomal abnormality rate of 5–10% among human spermatozoa (Brandiff *et al.*, 1986; Martin, 1985), gives a calculated incidence of chromosomal abnormalities among fertilized human oocytes after IVF of between 30.3 and 33.9% (Almeida and Bolton, 1994). In the only study to report the cytogenetic analysis of human pronucleate stage embryos, 65.2% of 46 embryos were found to be abnormal (Almeida and Bolton, 1996), and the reason for this discrepancy between the predicted and actual rates is not clear.

The same study found an overall abnormality rate during cleavage of 42.1% among 204 embryos analysed (Almeida and Bolton, 1996). This is in agreement with one other report (40%: Papadopoulos *et al.*, 1989), but compares with other reports of the incidence of chromosomal abnormalities among cleavage-stage embryos ranging from 12 (Zenzes and Casper, 1992) to 90% (Michaeli *et al.*, 1990). The wide variation in findings reflects the fact that the majority of analyses were undertaken using heterogeneous populations of embryos that varied in stage of development and degree of cytoplasmic fragmentation. Thus, while studies reporting the lowest incidence of chromosomal abnormalities were restricted to morphologically normal embryos (23%: Angell *et al.*, 1988) or populations of embryos where between 80 and 85% were morphologically normal (22.5%: Jamieson *et al.*, 1994; 29.2%: Plachot *et al.*, 1988), those studies reporting the highest incidence of abnormalities derived their results from populations including parthenogenetically activated, polypronucleate, developmentally delayed and arrested embryos (78.5%: Ma *et al.*, 1990; 74.5%: Wimmers and Van der Merwe, 1988), or embryos with poor morphology (90%: Michaeli *et al.*, 1990; Pellestor *et al.*, 1994). In human cleavage-stage embryos there is little doubt that chromosomal abnormalities are associated with developmental anomalies, including extracellular fragmentation (Almeida and Bolton, 1996; Munne *et al.*, 1995), multinucleated blastomeres (Kligman *et al.*, 1996), and developmental arrest (Munne *et al.*, 1995), in view of which the discrepancies between the findings of these studies are not surprising. The discrepancies can be further explained by the finding that there is a decline in the incidence of chromosomal abnormality between the pronucleate and 8-cell stages, from 65.2 to 27.4% (Almeida and Bolton, 1996).

In contrast to the plethora of studies investigating the incidence of chromosomal abnormalities in human oocytes and embryos, which utilize the relatively abundant resources of human oocytes that have failed to fertilize during therapeutic IVF, or embryos with poor morphology that are not suitable for cryopreservation, there have been very few studies investigating the origin of these abnormalities. In particular, it is interesting to determine whether their high incidence is an inherent characteristic of human reproduction, or whether it is an artefactual consequence of exposure to suboptimal conditions *in vitro*. One obvious potential source of damage to the cytoskeleton and, as a result, of disruption of chromosome distribution during therapeutic IVF is temperature fluctuation. The cytoskeleton of the human oocyte has been described (Pickering *et al.*, 1988) and a preliminary study demonstrated that the meiotic spindle suffers irreversible damage after cooling to room temperature (Pickering *et al.*, 1990). This finding was confirmed by a more extensive study, which found that cytoskeletal damage incurred by the human oocyte is rendered irreversible at some point between 2 and 10 min after exposure to room temperature, corresponding to a fall in temperature from 37°C to between 32°C and 27°C (Almeida and Bolton, 1995). Although no direct association was found between temperature-induced spindle anomalies and dispersal of chromosomes from the meiotic metaphase plate, parthenogenetic activation of temperature-shocked oocytes followed by cytogenetic analysis of cleavage-stage parthenotes demonstrated a direct association between temperature-induced spindle damage and the incidence of chromosomal abnormality in the parthenotes. Again, these effects were found to be rendered irreversible after between 2 and 10 min exposure to room temperature (Almeida and Bolton, 1995).

step of the RT–PCR, in the hope of increasing sensitivity (van Eijk et al., 1996). In contrast to the earlier findings, this study showed the presence of LIF-Rβ transcripts in human oocytes and throughout cleavage-stages of embryogenesis, and of gp130 in oocytes and 4-cell, but not later cleavage-stage embryos. The varying results obtained with different embryos at the same stage of development, such as the expression of gp130 in one blastocyst but not in two others (van Eijk et al., 1996) serve to emphasize the difficulties of studies using human embryos which show such variability, and of interpretation of findings. Nonetheless, it has been concluded that transcription of both components of the LIF-R takes place, and that functional LIF-R may be present throughout preimplantation human embryogenesis, suggesting that human embryos have the potential to respond to LIF secreted by the female reproductive tract.

6.2 Patterns of polypeptide synthesis

A single study has examined changes among the polypeptides already present in the human oocyte prior to fertilization during early embryogenesis (Capmany and Bolton, 1993). Using polyacrylamide gel electrophoresis (PAGE) and silver staining to detect proteins present at levels of 3–5 ng or greater, approximately 20 protein bands were resolved. Among these, the only differences identified were in a group of bands in the region of molecular weight 69 kDa, which resolved as one of two patterns in all the oocytes and embryos examined, up to the 8-cell stage. No relationship was identified between the presence of either pattern in the oocyte and its developmental potential, in terms of fertilization, pronucleus formation or cleavage. From this, it is concluded that any maternally inherited polypeptides that are crucial for the achievement of fertilization and/or development to the 8-cell stage are present at levels below 3–5 ng, and that any putative fertilization-associated protein must represent less than 3% of the total protein content of the oocyte, estimated to be 150 ng (Gifford et al., 1987).

A number of studies have examined the polypeptides synthesized by human oocytes and embryos, following incubation in ^{35}S-methionine and PAGE (Artley et al., 1992; Braude et al., 1988; Gifford et al., 1987). It is now clear that the marked changes that occur in the patterns of polypeptides synthesized by human embryos between the 4- and 8-cell stages are a reflection of major activation of the embryonic genome; in the presence of the transcriptional inhibitor α-amanitin, polypeptide synthesis and cleavage ceases at the 4-cell stage (Braude et al., 1988). As in other mammalian embryos, this transition from maternal to embryonic regulation of development coincides with the highest incidence of developmental arrest (Bolton et al., 1989; see review by Telford et al., 1990), and it has been suggested that the two are causally related (Braude et al., 1988, 1990). However, examination of the polypeptide synthetic profiles of human embryos that had undergone developmental arrest during cleavage has shown that cleavage arrest is not always associated with failure to synthesize embryonic proteins: of 54 embryos analysed, 27 showed evidence of transcription-related changes in patterns of polypeptide synthesis (Artley et al., 1992). However, 23 of these transcriptionally active embryos had arrested at the 4- to 8-cell stage, and only two each at the 3-cell and pronucleate stages. This compared with the group which failed to express embryonic proteins, which comprised primarily 2- to

3-cell embryos ($n = 17$), four pronucleate stage, and only five 4-cell and one 5-cell embryo. While this might suggest that chronological age, in combination with the number of completed cycles of cytokinesis, is involved in triggering embryonic gene activation, the same study reported a marked discrepancy between cell and nuclear number in both transcriptionally active and non-active arrested embryos (Artley *et al.*, 1992), leading to the suggestion that continuing karyokinesis is also important. A second study which compared the polypeptide synthetic patterns of abnormal and apparently normal human embryos reports somewhat different findings, with all 47 embryos that had arrested between the 2- and 8-cell stages expressing embryonic polypeptides, compared with only 7 of 13 embryos that had arrested at the pronucleate stage (G. Capmany and V.N. Bolton, unpublished data). The reason for the discrepancy between these two studies is unclear, but they both confirm that the relationship between cleavage arrest and activation of the embryonic genome is not straightforward.

A comparison of the polypeptide synthetic profiles of embryos that appear to be developing at a normal rate and showing good morphology with those of embryos showing poor morphology and/or slow cleavage or developmental arrest has been undertaken (G. Capmany and V.N. Bolton, unpublished data). Not surprisingly, even those embryos which appeared to be developing normally, with regular blastomeres and no fragmentation, showed differences in their patterns of polypeptide synthesis. However, the differences between embryos were progressively more marked with increasing morphological abnormality. Thus, arrested embryos with poor morphology display a larger number of, and greater variation among differences in polypeptides synthesized than embryos which are developing slowly. However, while increased abnormality of development appears to be reflected in increasingly abnormal patterns of polypeptide synthesis, it is not yet clear whether the differences in polypeptide synthetic profiles are the cause or result of the developmental anomalies.

6.3 Composition of the zona pellucida

The zona pellucida, the extracellular coat of glycoproteins which surrounds all mammalian oocytes, is the species-specific binding site for the spermatozoon, and is responsible for the development of the block to polyspermy. Its three component glycoproteins, ZP1, ZP2 and ZP3, and their modification following fertilization have been studied extensively in the mouse (Bleil *et al.*, 1988) but very little work has been undertaken in the human. Following radio-iodination of solubilized zonae, two components have been identified which, on the basis of their molecular weights are considered to correspond to mouse ZP2 and ZP3, but a glycoprotein corresponding to mouse ZP1 was not detected (Shabanowitz, 1990; Shabanowitz and O'Rand, 1988). Using the more sensitive biotinylation- and lectin-based assays and PAGE, three glycoprotein species corresponding to mouse ZP1–3 were identified (Moos *et al.*, 1995). Further investigation of the structural and functional properties of these zona glycoproteins has not been undertaken in the human.

6.4 Embryonic gene expression

Prior to the major activation of embryonic gene expression in the human, which appears to be between the 4- and 8-cell stages (Braude *et al.*, 1988; Tesarik *et al.*,

1988); see Section 6.2), development is regulated at the post-transcriptional level, utilizing stored maternal mRNA. In the mouse, it has been shown that following the transition from maternal to embryonic regulation of development, maternal mRNA is no longer detectable (Bolton *et al.*, 1984). With the advent of RT–PCR, it is now possible to look for transcripts of individual genes at different stages of development, and to examine patterns of gene expression, although few studies have addressed specifically the question of whether the transcripts identified are maternal or embryonic in origin. These will be considered in detail elsewhere (see Chapter 9).

The presence in the embryo of transcripts of the *SRY* gene, which encodes testis-determining factor and is on the Y chromosome, can only be a result of embryonic transcription. Using RT–PCR it has been demonstrated that *SRY* mRNA is present in human embryos throughout preimplantation development, from the 1-cell to the blastocyst stage (Fiddler *et al.*, 1995). Although this study used abnormal embryos that developed a single pronucleus or three pronuclei following IVF, it is unlikely that abnormal ploidy would have an effect on gene activation, and it can be concluded that some embryonic gene activity occurs in the human embryo prior to the major activation at the 4-cell stage.

An alternative approach to distinguishing between maternal and embryonic transcripts is the use of semiquantitation which is possible in conjunction with RT–PCR. An investigation of the patterns of expression of the cell cycle genes *c-mos* and cyclin-B1 found that levels of both cyclin B1 and β-actin mRNA in the embryo increase from the 6-cell stage onwards, indicating active transcription; in contrast, *c-mos* is expressed at relatively high levels in the oocyte, but its transcripts are barely detectable between the 6- and 16-cell stages (Heikinheimo *et al.*, 1995). These findings suggest that in the human as well as the mouse, the maternal pool of stored mRNA is degraded at the time embryonic transcription is activated.

7. Conclusion

The technique of IVF in the human has allowed direct investigation of aspects of preimplantation human embryogenesis. Although such studies are necessarily limited in comparison to those using animal models, they have, nonetheless, led to an increased understanding of human development. Further research, focusing on identifying the specific growth requirements of human embryos *in vitro*, and on mechanisms for distinguishing between viable and non-viable embryos, is necessary if the success rate of therapeutic IVF is to improve significantly.

References

Abramczuk J, Solter D, Koprowski A. (1977) The beneficial effect of EDTA on development of mouse one-cell embryos in chemically defined medium. *Dev. Biol.* **61:** 378–383.

Almagor M, Bejar C, Kafka I, Yaffe H. (1996) Pregnancy rates after communal growth of preimplantation human embryos in vitro. *Fertil. Steril.* **66:** 394–397.

Almeida PA, Bolton VN. (1993) Immaturity and chromosomal abnormalities in oocytes that fail to develop pronuclei following insemination *in vitro*. *Hum. Reprod.* **8:** 229–232.

Almeida PA, Bolton VN. (1994) The relationship between chromosomal abnormalities in the human oocyte and fertilization *in vitro*. *Hum. Reprod.* **9**: 343–346.

Almeida PA, Bolton VN. (1995) The effect of temperature fluctuations on the cytoskeletal organisation and chromosomal constitution of the human oocyte. *Zygote* **3**: 357–365.

Almeida PA, Bolton VN. (1996) The relationship between chromosomal abnormality in the human preimplantation embryo and development *in vitro*. *Reprod. Fertil. Dev.* **8**: 235–241.

Angell RR, Hillier SG, West JD, Glasier AF, Rodger MW, Baird DT. (1988) Chromosome anomalies in early human embryos. *J. Reprod. Fertil.* (Suppl.) **36**: 73–81.

Armstrong DT, Chaouat G, Guichard A, Cedard L, Andreu G, Denver L. (1989) Lack of a correlation of immunosuppressive activity secreted by human in vitro fertilized (IVF) ova with successful pregnancy. *J. In Vitro Fert. Embryo Transfer* **6**: 15–21.

Artley JK, Braude PR, Johnson MH. (1992) Gene activity and cleavage arrest in human pre-embryos. *Hum. Reprod.* **7**: 1014–1021.

Ashwood-Smith MJ, Hollands P, Edwards RG. (1989) The use of Albuminar 5 (TM) as a medium supplement in clinical IVF. *Hum. Reprod.* **4**: 702–705.

Austgulen R, Arntzen KJ, Vatten LJ, Kahn J, Sunde A. (1994) Detection of cytokines IL-1, IL-6, TGF-β and soluble TNF receptors in embryo culture fluids during *in vitro* fertilization. *Hum. Reprod.* **10**: 171–176.

Bavister BD. (1992) Co-culture for embryo development: is it really necessary? *Hum. Reprod.* **7**: 1339–1341.

Ben-Chetrit A, Jurisicova A, Casper RJ. (1996) Coculture with ovarian cancer cell enhances human blastocyst formation in vitro. *Fertil. Steril.* **65**: 664–666.

Biggers JD. (1987) Pioneering mammalian embryo culture. In: *The Mammalian Preimplantation Embryo* (ed. BD Bavister). Plenum Press, New York, pp. 1–21.

Birkenfeld A, Navot D. (1991) Endometrial cultures and their application to new reproductive technologies: a look ahead. *J. In Vitro Fert. Embryo Transfer* **8**: 119–126.

Bleil JD, Greve JM, Wassarman PM. (1988) Identification of a secondary sperm receptor in the mouse egg zona pellucida: role in maintenance of binding of acrosome-reacted sperm to eggs. *Dev. Biol.* **128**: 376–385.

Bolton VN. (1993) Implantation of human blastocysts following in vitro fertilization. In: *The Endocrinology of Embryo-Endometrium Interactions* (ed. S Glasser). Plenum Press, New York, NY, pp. 106–123.

Bolton VN, Oades PJ, Johnson MH. (1984) The relationship between cleavage, DNA replication, and activation of transcription in the mouse 2-cell embryo. *J. Embryol. Exp. Morphol.* **79**: 139–163.

Bolton VN, Hawes SM, Taylor CT, Parsons JH. (1989) Development of spare human preimplantation embryos *in vitro*: An analysis of the correlations among gross morphology, cleavage rates and development to the blastocyst. *J. In Vitro Fert. Embryo Transfer* **6**: 30–35.

Bolton VN, Wren ME, Parsons JH. (1991) Pregnancies after *in vitro* fertilization and transfer of human blastocysts. *Fertil. Steril.* **55**: 830–832.

Bongso A, Chye NS, Ratnam S, Sathananthan H, Wong PC. (1988) Chromosome anomalies in human oocytes failing to fertilize after insemination *in vitro*. *Hum. Reprod.* **3**: 645–649.

Bongso A, Soon-Chye N, Sathananthan H, Lian NP, Rauff M, Ratnam S. (1989) Improved quality of human embryos when co-cultured with human ampullary cells. *Hum. Reprod.* **4**: 706–713.

Borland RM. (1977) Transport processes in the mammalian blastocyst. In: *Development in Mammals, Vol. 1* (ed. MH Johnson). North-Holland, Amsterdam, pp. 31–67.

Borland RM, Biggers JD, Lechene CP, Taymor ML. (1980) Elemental composition of fluid in fallopian tube. *J. Reprod. Fertil.* **58**: 479–482.

Brandiff B, Gordon L, Ashworth LK, Watchmaker G, Carrano LV. (1986) Detection of chromosome abnormalities in human sperm. *Prog. Clin. Biol. Res.* **209B**: 469–476.

Braude P, Bolton V, Moore S. (1988) Human gene expression first occurs between the four- and eight-cell stages of preimplantation development. *Nature* **332**: 459–461.

Braude PR, Pickering SJ, Winston NJ, Artley JK, Johnson MH. (1990) The development of the human preimplantation embryo. In: *From Ovulation to Implantation; Proceedings of the III Renier de Graaf Symposium* (eds MJ Heineman, J Evers). Elsevier, Amsterdam, pp. 251–261.

Buster JE, Bustillo M, Rodi IA, Sydlee W, Cohen RNP, Hamilton M, Simon JA, Thorneycroft IH, Marshall JR. (1985) Biologic and morphologic development of donated human ova recovered by nonsurgical uterine lavage. *Am. J. Obstet. Gynecol.* **153**: 211–217.

Campbell S, Swann SR, Seif MW, Kimber SJ, Aplin JD. (1995) Cell adhesion molecules on the oocyte and preimplantation human embryo. *Mol. Hum. Reprod.* **1**, *Hum. Reprod.* **10**: 1571–1578.

Cansecoe RS, Sparks AE, Pearson RE, Gwazdauskas FC. (1992) Embryo density and medium volume effects on early murine embryo development. *J. Assist. Reprod. Genet.* **9**: 454–457.

Capmany G, Bolton VN. (1993) Polypeptide profiles of human oocytes and preimplantation embryos. *Hum. Reprod.* **8**: 1901–1905.

Capmany G, Taylor A, Braude PR, Bolton VN. (1996) The timing of pronuclear formation, DNA synthesis and cleavage in the human 1-cell embryo. *Molec. Hum. Reprod.* **2**: 299–306.

Chard T. (1991) Interferon-α is a reproductive hormone. *J. Endocrinol.* **131**: 337–338.

Chia CM, Winston RML, Handyside AH. (1995) EGF, TGF-α and EGF-R expression in human preimplantation embryos. *Development* **121**: 299–307.

Clark DA, Lee S, Fishel S *et al.* (1989) Immunosuppressive activity in human in vitro fertilization (IVF) culture supernatants and prediction of the outcome of embryo transfer: A multi-center trial. *J. In Vitro Fert. Embryo Transfer* **6**: 51–58.

Clarke RD. (1995) Assessing the viability of human preimplantation embryos. *Semin. Reprod. Endocrinol.* **13**: 53–63.

Cohen J, Wiemer K, Wright G. (1988) Prognostic value of morphological characteristics of cryopreserved embryo: A study using videocinematography. *Fertil. Steril.* **49**: 827–834.

Collier M, O'Neill C, Ammitt AJ, Saunders DM. (1988) Biochemical and pharmacological characterization of human embryo-derived platelet activating factor. *Hum. Reprod.* **8**: 993–999.

Collier M, O'Neill C, Ammitt AJ, Saunders DM. (1990) Measurement of human embryo-derived platelet-activating factor (PAF) using a quantitative bioassay of platelet aggregation. *Hum. Reprod.* **5**: 323–328.

Conaghan J, Handyside AH, Winston RML, Leese H. (1993) Effects of pyruvate and glucose on the development of human preimplantation embryos *in vitro*. *J. Reprod. Fertil.* **99**: 87–95.

Cummins JM, Breen TM, Fuller S, Harrison KL, Wilson LM, Hennessey JF, Shaw JM, Shaw G. (1986a) Comparison of two media in a human *in vitro* fertilization program: lack of significant differences in pregnancy rate. *J. In Vitro Fert. Embryo Transfer* **3**: 326–330.

Cummins JM, Breen TM, Harrison KL, Shaw JM, Wilson LM, Hennessey JF. (1986b) A formula for scoring human embryo growth rates in *in vitro* fertilization: Its value in predicting pregnancy and in comparison with visual estimates of embryo quality. *J. In Vitro Fert. Embryo Transfer* **3**: 284–295.

Dale B, Gualtieri R, Talevi R, Tosti E, Santella L, Elder K. (1991) Intercellular communication in the early human embryo. *Molec. Reprod. Dev.* **29**: 22–28.

Dale B, Tosti E, Iaccarino M. (1995) Is the plasma membrane of the human oocyte reorganised following fertilization and early cleavage? *Zygote* **3**: 31–36.

Dawson KJ, Rutherford AJ, Winston NJ, Shubak-Sharpe R, Winston RML. (1988) Human blastocyst transfer, is it a feasible proposition? *Hum. Reprod.* **3** (Suppl. 1): 44.

Daya S, Clark DA. (1986a) Production of immunosuppressive factors by preimplantation embryos. *Am. J. Reprod. Immunol. Microbiol.* **11**: 98–101.

Daya S, Clark DA. (1986b) Immunosuppressive factor(s) produced by in vitro fertilized human embryos. *N. Eng. J. Med.* **315**: 1551.

Dimitriadou F, Phocas I, Mantzavinos T, Sarandakou A, Rizos D, Zourlas PA. (1992) Discordant secretion of pregnancy specific β_1-glycoprotein and human chorionic gonadotropin by human pre-embryos cultured in vitro. *Fertil. Steril.* **57**: 631–636.

Djalali M, Rosenbusch B, Wolf M, Sterzik K. (1988) Cytogenetics of unfertilized human oocytes. *J. Reprod.* **84**: 647–652.

Dokras A, Sargent IL, Ross C, Gardner RL, Barlow DH. (1991) The human blastocyst: morphology and human chorionic gonadotrophin secretion *in vitro*. *Hum. Reprod.* **6**: 1143–1151.

Dokras A, Sargent IL, Barlow DG. (1993) Human blastocyst grading: an indicator of developmental potential? *Hum. Reprod.* **8**: 21–27.

Ducibella T, Albertini DF, Anderson E, Biggers JD. (1975) The preimplantation mammalian embryo: Characterization of intercellular junctions and their appearance during development. *Dev. Biol.* **45**: 231–250.

Dunglison GF, Barlow DG, Sargent IL. (1996) Leukaemia inhibitory factor significantly enhances the blastocyst formation rate of human embryos cultured in serum-free medium. *Hum. Reprod.* **11**: 191–196.

Eagle H. (1959) Amino acid metabolism in mammalian cell cultures. *Science* **130**: 432–437.

Edirisinghe WR, Murch AR, Yovich JL. (1992) Cytogenetic analysis of human oocytes and embryos in an in-vitro fertilization programme. *Hum. Reprod.* **7**: 230–236.

Edwards RG, Bavister BD, Steptoe PC. (1969) Early stages of fertilization *in vitro* of human oocytes matured *in vitro. Nature* **221**: 632–635.

Edwards RG, Steptoe PC, Purdy JM. (1970) Fertilization and cleavage *in vitro* of preovulatory human oocytes. *Nature* **227**: 1307–1309.

Fiddler M, Abdel-Rahman B, Rappolee DA, Pergament E. (1995) Expression of SRY transcripts in preimplantation human embryos. *Am. J. Med. Genet.* **55**: 80–84.

FitzGerald L, DiMattina M. (1992) An improved medium for long-term culture of human embryos overcomes the *in vitro* developmental block and increases blastocyst formation. *Fertil. Steril.* **57**: 641–647.

Fleming TP, George MA. (1987) Fluorescent latex microparticles: A non-invasive short-term cell lineage marker suitable for use in the mouse early embryo. *Roux. Arch. Dev. Biol.* **196**: 1–11.

Formigli L, Formigli G, Rocciao C. (1987) Donation of fertilized human ova to infertile women. *Fertil. Steril.* **47**: 162–165.

Freeman MR, Whitworth CM, Hill GA. (1995) Granulosa cell co-culture enhances human embryo development and pregnancy rate following in-vitro fertilization. *Hum. Reprod.* **10**: 408–414.

Fry RC, Batt PA, Fairclough RJ, Parr RA. (1992) Human leukemia inhibitory factor improves the viability of cultured ovine embryos. *Biol. Reprod.* **46**: 470–474.

Fusi FM, Vignalli M, Busacca M, Bronson RA. (1992) Evidence for the presence of an integrin on the oolemma of unfertilized human oocytes. *Mol. Reprod. Devel.* **31**: 215–222.

Gardner, Kaye. (1991) Insulin increases cell numbers and morphological development in mouse preimplantation embryos in vitro. *Reprod. Fentil. Devel.* **3**: 79–91.

Gardner DK, Sakkas D. (1993) Mouse embryo cleavage, metabolism and viability: role of medium composition. *Hum. Reprod.* **8**: 288–295.

Gifford DJ, Fleetham JA, Mahadevan MM, Taylor PJ, Schultz GA. (1987) Protein synthesis in mature human oocytes. *Gamete Res.* **18**: 97–107.

Giorgetti C, Terriou P, Auquier P, Hans E, Spach J-L, Salzmann J, Roulier R. (1995) Embryo score to predict implantation after in-vitro fertilization: based on 957 single embryo transfers. *Hum. Reprod.* **10**: 2427–2431.

Gott AL, Hardy K, Winston RLM, Leese HJ. (1990) Non-invasive measurement of pyruvate and glucose uptake and lactate production by single human preimplantation embryos. *Hum. Reprod.* **5**: 104–108.

Guidice LC. (1994) Growth factors and growth modulators in human uterine endometrium: their potential relevance to reproductive medicine. *Fertil. Steril.* **61**: 1–17.

Gunn LK, Homa ST, Searle MJ, Chard T. (1994) Lack of evidence for the production of interferon-α-like species by the cultured human pre-embryo. *Hum. Reprod.* **9**: 1522–1527.

Ham RG. (1965) Clonal growth of mammalian cells in a chemically defined, synthetic medium. *Proc. Natl Acad. Sci. USA* **53**: 288–293.

Handyside AH, Johnson MH. (1978) Temporal and spatial patterns of the synthesis of tissue-specific polypeptides in the preimplantation mouse embryo. *J. Embryol. Exp. Morphol.* **44**: 191–199.

Hardy K, Handyside AH, Winston RML. (1989a) The human blastocyst: cell number, death and allocation during late preimplantation development *in vitro. Development* **107**: 597–604.

Hardy K, Hooper MA, Handyside AH, Rutherford AJ, Winston RM, Leese HJ. (1989b) Non-invasive measurement of glucose and pyruvate uptake by individual human oocytes and preimplantation embryos. *Hum. Reprod.* **4**: 188–191.

Hardy K, Warner A, Winston RML. (1996) Expression of intercellular junctions during preimplantation development of the human embryo. *Mol. Hum. Reprod.* **2**: 621–632.

Harvey MB, Kaye PL. (1990) Insulin increases the cell number of the inner cell mass and stimulates morphological development of mouse blastocysts *in vitro. Development* **110**: 963–967.

Hawes S, Thomas M, Paton A, Farzaneh F, Bolton V. (1993) The detection of mRNA for the receptor of colony stimulation factor (*c-fms*) in human oocytes and preimplantation embryos. *J. Reprod. Fert.* (Abstr. Series) **12**: 7.

Heikinheimo O, Lanzendorf SE, Baka SG, Gibbons WE. (1995) Cell cycle genes *c-mos* and cyclin-B1 are expressed in a specific pattern in human oocytes and preimplantation embryos. *Mol. Hum. Reprod.* **10**: 699–707.

Hemmings R, Langlais J, Falcone T, Granger L, Miron P, Guyda H. (1992) Human embryos produce transforming growth factor alpha activity and insulin-like growth factor II. *Fertil. Steril.* **58**: 101–104.

Hertig AT, Rock J, Adams EC, Mulligan WJ. (1954) On the preimplantation stages of the human ovum: a description of four normal and four abnormal specimens ranging from the second to fifth day of development. *Contrib. Embryol.* **35**: 199–220.

HFEA (Human Fertilisation and Embryology Authority). (1996) *The Patients' Guide to DI and IVF Clinics.*

Holst N, Bertheussen K, Forsdahl F, Hakonsen MB, Hansen LJ, Nielsen HI. (1990) Optimization and simplification of culture conditions in human *in vitro* fertilization (IVF) and preembryo replacement by serum-free media. *J. In Vitro Fert. Embryo Transfer* **7**: 47–53.

Horne CHW, Towler CM, Pugh-Humphrey RGP, Thomson AQ, Bohn H. (1976) Pregnancy specific β_1-glycoprotein, a product of the syncytiotrophoblast. *Experientia* **32**: 1197–1199.

Jamieson ME, Coutts JRT, Connor JM. (1994) The chromosome constitution of human preimplantation embryos fertilized *in vitro*. *Hum. Reprod.* **9**: 709–715.

Johnson MH, Maro B. (1986) Time and space in the early mouse embryo: a cell biological approach to cell diversification. In: *Experimental Approaches to Mammalian Embryonic Development* (eds J Rossant, R Pedersen). Cambridge University Press, Cambridge, pp. 33–65.

Johnson MH, Eager D, Muggleton-Harris A. (1975) Mosaicism in organization of concanavalin A receptors on surface membrane of mouse oocytes. *Nature* **257**: 321–322.

Jones T, Ellison Z, Bolton V, Webley G, Milligan S. (1992) Bioassayable human chorionic gonadotrophin production by human preimplantation embryos. *J. Reprod. Fert.* (Abstr. Series) **9**: 58.

Jurisicova A, BenChetrit A, Varmuza SL, Casper RF. (1995) Recombinant human leukemia inhibitory factor does not enhance in vitro human blastocyst formation. *Fertil. Steril.* **64**: 999–1002.

Kligman I, Benadiva C, Alikani M, Munne S. (1996) The presence of multinucleated blastomeres in human embryos is correlated with chromosomal abnormalities. *Hum. Reprod.* **11**: 1492–1498.

Lane M, Gardner DK. (1992) Effect of incubation volume and embryo density on the development and viability of mouse embryos *in vitro*. *Hum. Reprod.* **7**: 558–562.

Larson RC, Ignotz GG, Currie WB. (1992) Platelet derived growth factor (PDGF) stimulates development of bovine embryos during the fourth cell cycle. *Development* **115**: 821–826.

Leese HJ. (1995) Metabolic control during preimplantation mammalian development. *Hum. Reprod. Update* **1**: 63–72.

Leese HJ, Barton AM. (1984) Pyruvate and glucose uptake by mouse ova and preimplantation embryos. *J. Reprod. Fertil.* **72**: 9–13.

Leese HJ, Hooper MAK, Edwards RG, Ashwood-Smith MJ. (1986) Uptake of pyruvate by early human embryos determined by a non-invasive technique. *Hum. Reprod.* **1**: 181–182.

Lopata A. (1992) The neglected human blastocyst. *J. In Vitro Fert. Embryo Transfer* **9**: 508–512.

Lopata A, Hay DL. (1989) The surplus human embryo: its potential for growth, blastulation, hatching, and human chorionic gonadotrophin production in culture. *Fertil. Steril.* **51**: 984–991.

Lopata A, Oliva K. (1993a) Regulation of chorionic gonadotropin secretion by cultured human blastocysts. In: *Preimplantation Embryo Development* (ed B Bavister). Springer-Verlag, New York, NY, pp. 276–295.

Lopata A, Oliva K. (1993b) Chorionic gonadotropin secretion by human blastocysts. *Hum. Reprod.* **8**: 932–938.

Ma S, Kalousek DK, Zouves C, Yuen BH, Gomel V, Moon YS. (1989) Chromosome analysis of human oocytes failing to fertilize *in vitro*. *Fertil. Steril.* **51**: 992–997.

Ma S, Kalousek DK, Zouves C, Yuen BH, Gomel V, Moon YS. (1990) The chromosomal complements of cleaved human embryos resulting from *in vitro* fertilization. *J. In Vitro Fert. Embryo Transfer* **7**: 16–21.

Magnuson T, Demsey A, Stackpole CW. (1977) Characterisation of intercellular junctions in the preimplantation mouse embryo by freeze fracture and thin section electron microscopy. *Dev. Biol.* **61**: 252–261.

Mansour RT, Aboulghar MR, Serour GI, Abbass A. (1994) Co-culture of human pronucleate oocytes with their cumulus cells. *Hum. Reprod.* **9**: 1727–1729.

Maro B, Gueth-Hallonet C, Aghion J, Antony C. (1991) Cell polarity and microtubule organization during mouse early embryogenesis. *Development* (Suppl. 1), 17–25.

Marquant-LeGuinne B, Humbolt P, Guillon N, Thibier M. (1993) Murine LIF improves the development of IVF cultured bovine morulae. *J. Reprod. Fertil.* (Abstract) **12:** 61.

Martin RH. (1985) Chromosomal abnormalities of human sperm. In: *Aneuploidy: Aetiology and Mechanisms* (eds VL Dellarco, PE Voytek, A Hollaender). Plenum Press, New York, NY, pp. 91–102.

Martin RH, Mahadevan MM, Taylor PJ, Hildebrand K, Long-Simpson L, Peterson D, Tamamoto J, Fleetham J. (1986) Chromosomal analysis of unfertilized human oocytes. *J. Reprod. Fertil.* **78:** 673–678.

Menezo YJR, Testart J, Perrone D. (1984) Serum is not necessary in human *in vitro* fertilization, early embryo culture, and transfer. *Fertil. Steril.* **42:** 750–755.

Menezo YJR, Guerin J-F, Czyba J-C. (1990) Improvement of human early embryo development *in vitro* by coculture on monolayers of Vero cells. *Biol. Reprod.* **42:** 301–306.

Menezo Y, Hazout A, Dumont M, Herbaut N, Nicollet B. (1992) Coculture of embryos on Vero cells and transfer of blastocysts in humans. *Hum. Reprod.* **7:** 101–106.

Michaeli G, Fejgin M, Ghetler Y, Ben Nun I, Beyth Y, Amiel A. (1990) Chromosomal analysis of unfertilized oocytes and morphologically abnormal preimplantation embryos from an *in vitro* fertilization program. *J. In Vitro Fert. Embryo Transfer* **7:** 341–346.

Moessner J, Dodson WC. (1995) The quality of human embryo growth is improved when embryos are cultured in groups rather than separately. *Fertil. Steril.* **64:** 1034–1035.

Moos J, Faundes D, Kopf GS, Schultz RM. (1995) Composition of the human zona pellucida and modifications following fertilization *Hum. Reprod.* **10:** 2467–2471.

Mottla GL, Adelman MR, Hall JL, Gindoff PR, Stillman RJ, Johnson KE. (1995) Lineage tracing demonstrates that blastomeres of early cleavage-stage human pre-embryos contribute to both trophectoderm and inner cell mass. *Hum. Reprod.* **10:** 384–391.

Mroz EA, Lechene CP. (1980) Fluorescence analysis of picolitre samples. *Anal. Biochem.* **102:** 90–96.

Muggleton-Harris AL, Findlay I, Whittingham DG. (1990) Improvement of the culture conditions for the development of human preimplantation embryos. *Hum. Reprod.* **5:** 217–220.

Munne S, Alikani M, Tomkin G, Grifo J, Cohen J. (1995) Embryo morphology, developmental rates, and maternal age are correlated with chromosome abnormalities. *Fertil. Steril.* **64:** 382–391.

Nadijcka M, Hillman N. (1974) Ultrastructural studies of the mouse blastocyst substages. *J. Embryol. Exp. Morphol.* **32:** 675–695.

Nakatsuka M, Yoshida N, Kudo T. (1992) Platelet activating factor in culture media as an indicator of human embryonic development after in-vitro fertilization. *Hum. Reprod.* **7:** 1435–1439.

Nonogaki T, Noda Y, Narimoto K, Umaoka Y, Mori T. (1991) Protection from oxidative stress by thioredoxin and superoxide dismutase of mouse oocytes fertilized *in vitro. Hum. Reprod.* **6:** 1305–1310.

Olivennes F, Hazout A, Lelaidier C, Freitas S, Fanchin R, de Ziegler D, Frydman R. (1994) Four indications for embryo transfer at the blastocyst stage. *Hum. Reprod.* **9:** 2367–2373.

O'Neill C. (1985) Examination of the causes of early thrombocytopaenia in mice. *J. Reprod. Fertil.* **73:** 578–585.

O'Neill C, Gidley-Baird AA, Pike IL, Saunders DM. (1987) Use of a bio-assay for embryo-derived platelet-activating factor as a means of assessing quality and pregnancy potential of human embryos. *Fertil. Steril.* **47:** 969–975.

Papadopoulos G, Templeton AA, Fisk N, Randell J. (1989) The frequency of chromosome anomalies in human preimplantation embryos after *in vitro* fertilization. *Hum. Reprod.* **4:** 91–98.

Paria BC, Dey SK. (1990) Preimplantation embryo development *in vitro*: cooperative interactions among embryos and role of growth factors. *Proc. Natl Acad. Sci. USA* **87:** 4756–4760.

Pellestor F, Sele B. (1988) Assessment of aneuploidy in the human female by using cytogenetics of IVF failures. *Am. J. Hum. Genet.* **42:** 274–283.

Pellestor F, Dufour MC, Arnal F, Humeau C. (1994) Direct assessment of the fate of chromosomal abnormalities in grade IV human embryos produced by *in vitro* fertilization procedure. *Hum. Reprod.* **9:** 293–302.

Phocas I, Sarandakou A, Rizos D, Dimitriadou F, Mantzavinos T, Zourlas PA. (1992) Secretion of alpha-immunoreactive inhibin by human pre-embryos cultured in vitro. *Hum. Reprod.* **7:** 545–549.

Pickering SJ, Johnson MH, Braude PR, Houliston E. (1988) Cytoskeletal organisation in fresh, aged and spontaneously activated human oocytes. *Hum. Reprod.* **3:** 978–989.

Pickering SJ, Cant A, Braude PR, Currie J, Johnson MH. (1990) Transient cooling to room temperature can cause irreversible disruption of the meiotic spindle in the human oocyte. *Fertil. Steril.* **54**: 102–108.

Pieters MHEC, Geraedts JPM, Dumoulin JCM, Evers JLH, Bras M, Kornips FHJAC, Menheere PPCA. (1989) Cytogenetic analysis of *in vitro* fertilization (IVF) failures. *Hum. Genet.* **81**: 367–370.

Plachot M, Veiga A, Montagut J *et al.* (1988) Are clinical and biological IVF parameters correlated with chromosomal disorders in early life: a multicentric study. *Hum. Reprod.* **3**: 627–635.

Plachot M, Heymann D, Godard A, Antoine JM, Salat-Baroux J. (1995) Embryo coculture: the interaction embryo-feeder. In: *IXth World Congress on In Vitro Fertilization and Assisted Reproduction* (eds A Abarumieh, E Bernat *et al.*). Monduzzi, Editore, Bologna, pp. 73–97.

Quinn P, Margalit R. (1996) Beneficial effects of coculture with cumulus cells on blastocyst formation in a prospective trial with supernumerary human embryos. *J. Assist. Reprod. Genet.* **13**: 9–14.

Quinn P, Kerin JF, Warnes GM. (1985) Improved pregnancy rate in human *in vitro* fertilization with the use of a medium based on the composition of human tubal fluid. *Fertil. Steril.* **44**: 493–498.

Robertson SA, Seamark RF, Guilbert LJ, Wegmann TG. (1994) The role of cytokines in gestation. *Crit. Rev. Immunol.* **14**: 239–292.

Ryan JP, O'Neill C. (1994) Embryo-derived and maternal factors associated with developmental potential and viability of preimplantation mammalian embryo development. In: *The Biological Basis of Early Reproductive Failure: Applications to Medically-Assisted Conception* (ed. J van Blerkom). Oxford University Press, Oxford, pp. 374–405.

Saith RR, Bersinger NA, Barlow DH, Sargent IL. (1996) The role of pregnancy-specific β-1 glycoprotein (SP1) in assessing human blastocyst quality *in vitro*. *Hum. Reprod.* **11**: 1038–1042.

Sakkas D, Jaquenoud N, Leppens G, Campana A. (1994) Comparison of results after in vitro fertilized human embryos are cultured in routine medium and in coculture on Vero cells: A randomised study. *Fertil. Steril.* **61**: 521–525.

Santella L, Alikani M, Talansky B, Cohen J, Dale B. (1992) Is the human plasma membrane polarised? *Hum. Reprod.* **7**: 999–1003.

Sathananthan H, Bongso A, Ng S-C, Ho J, Mok H, Ratnam S. (1990) Ultrastructure of preimplantation human embryos co-cultured with human ampullary cells. *Hum. Reprod.* **5**: 309–318.

Schmiady H, Kentenich H. (1989) Premature chromosome condensation after *in vitro* fertilization. *Hum. Reprod.* **4**: 689–695.

Segars JH, Rogers BJ, Niblack GD, Osteen KG, Wentz AC. (1989) The human blastocyst produces a soluble factor(s) that interferes with lymphocyte proliferation. *Fertil. Steril.* **52**: 381–387.

Shabanowitz RB. (1990) Mouse antibodies to human zona pellucida: evidence that human ZP3 is strongly immunogenic and contains two distinct isomer chains. *Biol. Reprod.* **43**: 260–270.

Shabanowitz RB, O'Rand MG. (1988) Characterization of the human zona pellucida from fertilized and unfertilized eggs. *J. Reprod. Fertil.* **82**: 151–161.

Sharkey A. (1995) Cytokines and embryo/endometrial interactions. *Reprod. Med. Rev.* **4**: 87–100.

Sharkey AM, Dellow K, Blayney M, Macnamee M, Charnock-Jones S, Smith SK. (1995) Stage-specific expression of cytokine and receptor messenger ribonucleic acids in human preimplantation embryos. *Biol. Reprod.* **53**: 955–962.

Sheth KV, Roac GL, Al-Sedairy ST, Parhar RS, Hamilton CJCM, Al-Abdul Jabbar F. (1991) Prediction of successful embryo implantation by measuring interleukin-1 alpha and immunosuppressive factors in preimplantation embryo culture fluid. *Fertil. Steril.* **55**: 952–957.

Shulman A, Ben-Nun I, Ghetler Y, Kaneti H, Shilon M, Beyth Y. (1993) Relationship between embryo morphology and implantation rate after in vitro fertilization treatment in conception cycles. *Fertil. Steril.* **60**: 123–126.

Smotrich DB, Stillman RJ, Widra EA, Gindoff PR, Kaplan P, Graubert M, Johnson KE. (1996) Immunocytochemical localization of growth factors and receptors in human pre-embryos and Fallopian tubes. *Hum. Reprod.* **11**: 184–190.

Svalander PC, Holmes PV, Olovsson M, Wikland M, Gemzell Danielsson K, Bygdeman M. (1991) Platelet-derived growth factor is detected in human blastocyst culture medium but not in human follicular fluid — a preliminary report. *Fertil. Steril.* **56**: 367–369.

Tarin JJ, Gomez E, Pelleir A. (1991) Chromosome anomalies in human oocytes *in vitro*. *Fertil. Steril.* **55**: 964–969.

Telford N, Watson A, Schultz G. (1990) Transition from maternal to embryonic control in early mammalian development: A comparison of several species. *Mol. Reprod. Dev.* **26:** 90–100.

Tesarik J, Kopecny V, Plachot M, Mandelbaum J. (1988) Early morphological signs of embryonic genome expression in human preimplantation development as revealed by quantitative electron microscopy. *Dev. Biol.* **128:** 15–20.

Testart J. (1986) cleavage-stage of human embryos two days after fertilization *in vitro* and their developmental ability after transfer into the uterus. *Hum. Reprod.* **1:** 29–31.

Thibodeaux JK, Godke RA. (1995) Potential use of embryo coculture with human in vitro fertilization procedures. *J. Assist. Reprod. Genet.* **12:** 665–677.

Turner K, Lenton EA. (1996) The influence of Vero cell culture on human embryos development and chorionic gonadotrophin production *in vitro. Hum. Reprod.* **11:** 1966–1974.

Turpeenniemi-Hujanen T, Ronnberg L, Kauppila A, Puistola U. (1992) Laminin in the human embryo implantation: analogy to the invasion by malignant cells. *Fertil. Steril.* **58:** 105–113.

Turpeenniemi-Hujanen T, Feinberg RF, Kauppila A, Puistola U. (1995) Exracellular matrix interactions in early human embryos: implications for normal implantation events. *Fertil. Steril.* **64:** 132–138.

Van Blerkom J. (1993) Development of human embryos to the hatched blastocyst stage in the presence or absence of a monolayer of Vero cells. *Hum. Reprod.* **8:** 1525–1539.

van Eijk MJT, Mandelbaum J, Salat-Baroux J, Belaisch-Allart J, Plachot M, Junca AM, Mummery CL. (1996) Expression of leukaemia inhibitory factor receptor subunits LIFRβ and gp130 in human oocytes and preimplantation embryos. *Molec. Hum. Reprod.* **2:** 355–360.

Veiga A, Claderon G, Santalo J, Barri PN, Egozcue J. (1987) Chromosome studies in oocytes and zygotes from an IVF programme. *Hum. Reprod.* **2:** 425–430.

Vlad M, Walker D, Kennedy RC. (1996) Nuclei number in human embryos co-cultured with human ampullary cells. *Hum. Reprod.* **11:** 1678–1686.

Wiemer KE, Cohen J, Amborski GF, Wright G, Wiker SR, Munyakazi L, Godke RA. (1989a) In-vitro development and implantation of human embryos following culture on fetal bovine uterine fibroblast cells. *Hum. Reprod.* **4:** 595–600.

Wiemer KE, Cohen J, Wiker SR, Malter HE, Wright G, Godke RA. (1989b) Coculture of human zygotes on fetal bovine uterine fibroblasts: embryonic morphology and implantation. *Fertil. Steril.* **52:** 503–508.

Wiemer KE, Hoffman DI, Maxson WS, Eager S, Muhlberger B, Fiore I, Cuervo M. (1993) Embryonic morphology and rate of implantation of human embryos following co-culture on bovine oviductal epithelial cells. *Hum. Reprod.* **8:** 97–101.

Wiker S, Malter H, Wright G, Cohen J. (1990) Recognition of paternal pronuclei in human zygotes. *J. In Vitro Fert. Embryo Transfer* **7:** 33–37.

Wimmers MSE, Van der Merwe JV. (1988) Chromosome studies on early human embryos fertilized *in vitro. Hum. Reprod.* **3:** 894–900.

Winston NJ, Braude PR, Pickering SJ, George MA, Cant A, Currie J, Johnson MH. (1991) The incidence of abnormal morphology and nucleocytoplasmic ratios in 2-, 3- and 5-day human pre-embryos. *Hum. Reprod.* **6:** 17–24.

Witkin S, Liu H-C, Davis OK, Rosenwaks Z. (1991) TNF is present in maternal sera in embryo culture fluids during IVF. *J. Reprod. Immunol.* **19:** 85–93.

Wood SA, Kaye PL. (1989) Effects of epidermal growth factor on preimplantation mouse embryos. *J. Reprod. Fertil.* **85:** 575–582.

Woodward BJ, Lenton EA, Turner K. (1993) Human chorionic gonadotrophin : embryonic secretion is a time dependent phenomenon. *Hum. Reprod.* **8:** 1463–1468.

Woodward BJ, Lenton EA, Turner K. (1994) Embryonic HCG secretion and hatching: poor correlation with cleavage rate and morphological assessment during preimplantation development *in vitro. Hum. Reprod.* **9:** 1909–1914.

Wright G, Wiker S, Elsner C, Kort H, Massey J, Mitchell D, Toledo A, Cohen J. (1990) Observations on the morphology of pronuclei and nucleoli in human zygotes and implications for cryopreservation. *Hum. Reprod.* **5:** 109–115.

Zenzes MT, Casper RF. (1992) Cytogenetics of human oocytes, zygotes and embryos after *in vitro* fertilization. *Hum. Genet.* **88:** 367–375.

Zolti M, Ben-Rafael Z, Meironi K, Shemesh M, Bider D, Maschiach S. (1991) Cytokine involvement in oocytes and early embryos. *Fertil. Steril.* **56:** 265–272.

9

Gene expression in human preimplantation embryos

Rob Daniels and Marilyn Monk

1. Introduction

Research towards a greater understanding of early human development is limited by the scarcity and precious nature of embryonic material available. There are also ethical limitations on the use of human embryos for research, and rightly so. To date, the vast majority of research investigating gene activity in mammalian development has been carried out in the mouse (for reviews see Kidder, 1992; Nothias *et al.*, 1995). However, the few studies that have been carried out in the human have revealed developmental timing differences between human and the mouse (Telford *et al.*, 1990) and thus, although mouse models are helpful in the first instance, it is important finally to study human embryos directly. The purpose of molecular analyses of human development is, at this early stage, largely academic. Nevertheless, we can anticipate that an increased understanding of the timing of onset and the tissue-specificity of expression of specific genes, and their role during human preimplantation development, will lead to improved *in vitro* fertilization (IVF) procedures and new approaches to contraception. In addition, determination of the timing and tissue-specificity of gene transcription gives information about the onset of a genetic disease (a mutation in a gene is only manifested when and where that gene is expressed). Finally, the new single-cell-sensitive procedures developed for these studies will increase the range of genetic diseases which may be diagnosed very early in development, for example, by preimplantation diagnosis, thus avoiding a more traumatic abortion later in pregnancy.

The spread of IVF techniques has resulted in an increase in the number of human preimplantation embryos available for research, although such material is still very rare and it is necessary to obtain the maximum information possible from each individual embryo. To this end, the single cell procedures must be shown to be efficient and absolutely reliable using, in the first instance, human somatic cells, such as lymphocytes or fibroblasts. Even so, analyses of transcription of specific

Molecular Genetics of Early Human Development, edited by T. Strachan, S. Lindsay and D.I. Wilson.
© 1997 BIOS Scientific Publishers Ltd, Oxford.

genes at such early stages of development must be interpreted with caution; the results obtained and conclusions drawn are subject to the limitations of the sensitivity of the techniques used. In this chapter, we review current research on gene transcription in human preimplantation development and assess the significance of the results obtained.

2. Onset of gene transcription in the human zygote

During oogenesis, a supply of maternal proteins and mRNA is produced which, following fertilization, supports the initial development of the mammalian zygote. As development proceeds, these maternally inherited proteins and mRNAs are degraded and replaced by those produced by the zygote itself. Several studies have been carried out to determine when the 'switch' from maternal to embryonic control of development occurs (Braude *et al.*, 1988; Tesarik *et al.*, 1986). However, this transfer of genetic control is likely to happen over a transitional period, with the onset of embryonic transcription overlapping with the period when maternal transcripts are still present. Studies involving the incubation of human preimplantation embryos from the 1-cell stage in a medium containing α-amanitin, which inhibits transcription, have shown that development of human embryos may proceed to the 4- to 8-cell stage, presumably supported solely by maternal transcripts. It is assumed that beyond this stage embryonic transcripts are required for further development (Braude *et al.*, 1988) and, indeed, initial studies monitoring the onset of gene expression in human preimplantation embryos, carried out by autoradiography (Tesarik *et al.*, 1986) or protein analysis (Braude *et al.*, 1988), detect synthesis of new transcription and new proteins already present in the 4-cell human zygote.

Similar experiments on preimplantation embryos of the mouse, cow, sheep and pig have revealed species differences in that preimplantation development in these species is blocked at the 2-cell, 8- to 12-cell, 8- to 16-cell and 4-cell stages, respectively, following incubation from the 1-cell stage in a medium containing α-amanitin (Telford *et al.*, 1990). However, although these studies provide important information as to the onset of transcription in the mammalian zygote and identify species differences at this early stage of development, it is probable that development is under the influence of transcripts from the zygotic genome at a much earlier stage than these studies suggest.

When specific enzymes are analysed by their activity on a given substrate, it is not always easy to distinguish between maternally inherited enzyme and that produced by the zygote itself (Braude *et al.*, 1989; Sermon *et al.*, 1992; West *et al.*, 1989). Again, species differences emerge; for example, the first model system to demonstrate the feasibility of preimplantation diagnosis of genetic disease was developed in a mouse model for Lesch–Nyhan disease (Monk *et al.*, 1987). Male embryos deficient for the X-linked enzyme hypoxanthine phosphoribosyl transferase (hprt) could be easily identified by direct sensitive enzyme assay. However, when the same enzyme assay was applied to human preimplantation embryos it was immediately apparent that new *HPRT* enzyme activity due to zygotic transcription of the *HPRT* gene, was masked by the relatively high level of *HPRT* enzyme activity inherited in the human egg cytoplasm (Braude *et al.*, 1989). In other words, whereas this diagnostic technique worked in the mouse, direct enzyme assay could not be used to diagnose human male preimplantation embryos carrying the X-linked Lesch–Nyhan disease mutation. These results highlight the need finally to carry out research on the human embryos themselves and not to extrapolate from other mammalian systems (although such model systems are invaluable in the first instance).

A recent major technical advance in the study of gene expression in preimplantation embryonic development is the reverse transcriptase polymerase chain reaction (RT–PCR), which is capable of monitoring the presence of specific mRNAs in a single cell (*Figure 1*). By designing the RT–PCR procedure to span known common polymorphisms in the gene of interest, it is possible to distinguish the maternal and paternal alleles. We cannot distinguish between maternal transcripts which are inherited in the egg cytoplasm from those due to new transcription from the maternally inherited allele in the embryo. However, the presence of paternal transcripts indicates that the zygotic genome is active.

2.1 Some tissue-specific genes are expressed in the human preimplantation embryo

The first RT–PCR studies on gene transcription in human preimplantation development showed that zygotic gene transcription could occur immediately after fertilization (Ao *et al.*, 1994; Daniels *et al.*, 1995; Fiddler *et al.*, 1995), much earlier than previously reported using different methods (Braude *et al.*, 1988; Tesarik *et al.*, 1986). It is surprising however, that genes which are thought to be tissue-specific, and which are known to function at later stages of development, are being expressed so early. We have examined the expression of the myotonin protein kinase (MPK) gene in individual human preimplantation embryos. MPK contains a highly polymorphic triplet repeat sequence in its 3' untranslated region which allows us to distinguish the different mRNAs produced by the two parental alleles by the size of their amplified cDNAs on an agarose gel (*Figure 2*; from Daniels *et al.*, 1995). The maternal and paternal genotypes could be determined from cumulus cells and sperm, respectively. The onset of transcription of the *MPK* gene can be seen by the presence of transcripts of the paternally inherited *MPK* allele already in embryos at the 1-cell stage. Similarly, transcripts from paternally inherited genes on the Y chromosome, *SRY* and *ZFY*, have been detected in the 1-cell human zygote (Ao *et al.*, 1994; Fiddler *et al.*, 1995). Using RT–PCR, transcripts of MHC class I genes have also been shown in the mouse 1-cell zygote (Sprinks *et al.*, 1993).

The significance of zygotic gene transcription at this early developmental stage is not clear. MPK, although expressed at low levels in a number of tissues, is predominantly a muscle-specific protein (Salvatori *et al.*, 1994; Whiting *et al.*, 1995). An expansion of the triplet repeat in the *MPK* gene, possibly affecting gene transcription (Fu *et al.*, 1993; Sabouri *et al.*, 1993), has been implicated in the phenotype of the disease myotonic dystrophy (Brook *et al.*, 1992). Since *MPK* transcription occurs so early in preimplantation embryos, one might be tempted to conclude that embryos with large repeat expansions would be affected as early

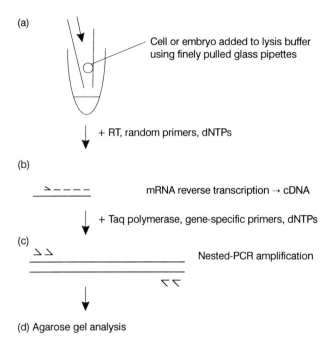

(a)

Cell or embryo added to lysis buffer
using finely pulled glass pipettes

+ RT, random primers, dNTPs

(b)

mRNA reverse transcription → cDNA

+ Taq polymerase, gene-specific primers, dNTPs

(c)

Nested-PCR amplification

(d) Agarose gel analysis

Figure 1. Reverse transcriptase polymerase chain reaction (RT–PCR) assay of gene
expression in a single cell or embryo. (a) The cell or embryo is lysed by heating to 80°C in
lysis buffer, and (b) the mRNA reverse transcribed by the addition of reverse
transcriptase (RT), random hexamer primers (⌐) and, the four deoxynucleotide
triphosphates (dNTPs) to produce complementary DNA (cDNA). In some cases, to
improve the efficiency and specificity of the RT–PCR, reverse transcription may be
primed using a gene-specific primer. In addition, reverse transcription may be primed at
the poly A of the mRNA tail by use of an oligo (dT) primer; however, this limits the
cDNAs produced to the 3′ region of the mRNA. (c) Following reverse transcription, the
cDNA is amplified by a nested or hemi-nested PCR. (d) The PCR amplification products
are visualized on an ethidium-bromide-stained agarose gel. For more detailed
methodology see Daniels *et al.* (1995, 1997a,b). Primers (⌐) for PCR amplification are
designed to span one or more introns to allow genomic DNA and complementary DNA
products to be distinguished by size on the agarose gel. In addition, transcripts from a
number of genes can be analysed in a single cell either by multiplex PCR or by
amplifying small aliquots of the reverse transcription products separately with different
gene-specific primers. In preference, the PCR sequence amplified will encompass a
known DNA sequence polymorphism in order to distinguish transcripts from maternally
and paternally inherited alleles either by PCR product size (Daniels *et al.*, 1995),
restriction enzyme digest or by using the single nucleotide primer extension assay
(SNuPE, Buzin *et al.*, 1994). Prior to application of the RT–PCR techniques to
preimplantation embryos, it is necessary to assess the efficiency and sensitivity of the
procedure on single somatic cells known to express the gene under study.

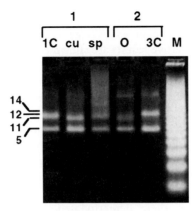

Figure 2. Detection of transcripts of myotonin protein kinase (*MPK*) by reverse transcription polymerase chain reaction (RT–PCR) in a 1-cell embryo (1C), 3-cell embryo (3C) and an unfertilized oocyte (O). Maternal and paternal genotypes of the embryo from family one, obtained by PCR from cumulus cells (cu) and sperm (sp), respectively, are also shown. M – size marker (1 kb DNA ladder). The number of CTG triplet repeats contained in the different alleles (estimated from the size of the amplified sequence) is indicated. For family 1, the 12-repeat allele expressed in the 1-cell embryo subjected to RT–PCR corresponds to the larger of the two paternal alleles detected by PCR in sperm, and not to either of the two smaller maternal alleles (5 repeat and 11 repeat) detected by PCR of cumulus cells. In the 3-cell embryo, for family 2, both maternal alleles are transcribed and detected by RT–PCR in the unfertilized oocyte. Transcripts of the 14-repeat allele detected in the 3-cell embryo must, therefore, be derived from a paternally inherited allele. The derivation of the transcripts detected from the paternally inherited allele in these embryos confirms new embryonic transcription of the *MPK* gene in the zygote. Reproduced from Daniels *et al.* (1995) by permission of Oxford University Press.

as the 1-cell stage. The *SRY* gene encodes the Y-linked testis-determining factor (Clepet *et al.*, 1993), and the *ZFY* gene maps to a region essential for spermatogenesis (Koopman *et al.*, 1991). Both genes function later in development at the time of testis formation (Clepet *et al.*, 1993; Koopman *et al.*, 1991). It is difficult to imagine that the expression of these tissue-specific genes has functional significance in preimplantation embryos.

Alternatively, it seems probable that very early tissue-specific gene expression has no specific function but is a consequence of some aspect of fertilization. We have investigated the possibility that a generalized derepression of the paternal genome occurs immediately after fertilization, perhaps as a consequence of 'swelling' of the male pronucleus or of the release of protamines associated with sperm heterochromatin, resulting in the 'illegitimate' transcription of paternally inherited genes. If this were the case, we might expect to see the presence of transcripts of other tissue-specific genes, which would not be expected to have a function in early development. We therefore developed RT–PCR procedures sensitive enough to detect a single β-globin molecule to look for transcription of this gene in human preimplantation embryos. However, we were unable to detect transcripts of the β-globin gene in the 1-cell human zygote and a positive signal for β-globin transcription was detected in only one of the eleven later-stage human embryos analysed. This is in contrast to the level of transcription of the ubiquitously expressed house-keeping gene, *HPRT*, which was detected in eight of the embryos analysed. Representative results are shown in *Figure 3*.

A second possibility to be considered is that a derepression of transcriptional control occurs for paternally inherited genes containing CpG islands (*MPK*, *ZFY* and *SRY* all have associated CpG islands, whereas the β-globin does not). Genes which show tissue-specific expression and contain CpG islands are known to be hypomethylated except for imprinted genes or genes on the inactive X chromosome in females (reviewed in Monk, 1986). In mature sperm, CpG islands are known to be hypomethylated (Zhang *et al.*, 1987), although, due to lack of cytoplasm, transcription does not normally occur. Following fertilization, paternally inherited genes containing CpG islands may be potentially transcriptionally active due to their demethylated status and the favourable environment of the oocyte. Therefore, as a control for the non-island β-globin gene, which is not expressed in the zygote, we have analysed preimplantation embryos for the presence of transcripts of the α-globin gene, a gene which contains a CpG island and shows tissue-specific expression. In contrast to β-globin, transcripts of α-globin are detected in six out of nine preimplantation embryos tested from the 1-cell to the 8-cell stage of development. Representative results showing α-globin and *HPRT* transcripts in a 1-cell and a 4-cell embryo are shown in *Figure 4*. The detection of α-globin transcripts in human preimplantation embryos suggests that a derepression of transcriptional activity may occur for genes containing CpG islands in the initial stages of human preimplantation development. This hypothesis is supported by the inability to detect transcripts of the IGF1 insulin genes (genes which show tissue-specific expression but do

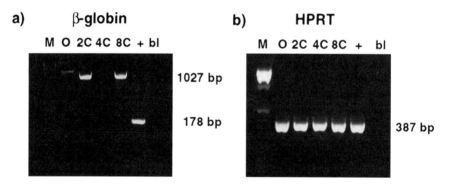

Figure 3. Analysis of β-globin and *HPRT* expression in individual 2-cell (2C), 4-cell (4C) and 8-cell (8C) human preimplantation embryos (Daniels *et al.*, 1997a). The positive controls (+) are single red blood cells for β-globin transcription and single fibroblasts for *HPRT*. Lanes marked bl are blank polymerase chain reactions (PCR) (without the addition of cDNA). The 1027 bp band for β-globin, detected at all stages analysed, is the product of genomic DNA amplification by the PCR reaction which is sensitive to a single molecule of β-globin sequence. The 178 bp β-globin cDNA product is not detected in any of the embryos shown. The detection of *HPRT* transcripts in the same embryo samples confirms the functioning of the transcription machinery in these embryos. M, size marker (1 kb DNA ladder).

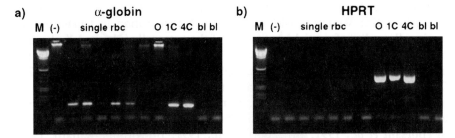

Figure 4. Analysis of α-globin and *HPRT* expression in an unfertilized oocyte (O), and in individual 1-cell (1C) and 4-cell (4C) embryos (Daniels *et al.*, 1997a). Transcripts of α-globin are detected in the 1-cell and 4-cell embryos but not in the unfertilized oocyte. *HPRT* transcripts are detected in the same oocyte, and 1-cell and 4-cell embryos. The detection of α-globin in the 1-cell embryo and its absence in the unfertilized oocyte show that the onset of embryonic transcription for α-globin occurs very soon after fertilization The lane marked (-) indicates a single red blood cell (rbc) negative control (no reverse transcriptase added). Lanes marked bl are blank polymerase chain reactions (PCR) (without the addition of cDNA). M, size marker (1 kb DNA ladder). Note that single red blood cells show the presence of α-globin transcripts but the absence of *HPRT* transcripts, as expected.

not contain a CpG island) in human preimplantation embryos (Lighten *et al.*, 1997), and by the detection of transcripts in the 1-cell human zygote from the *MPK*, *SRY* and *ZFY* genes, each of which contains a CpG island and shows tissue-specific expression.

3. X-Chromosome inactivation

In female mammals, one of the two X chromosomes is inactivated to compensate for the extra dosage of X-chromosome-linked genes present in female mammals with respect to that in males (Lyon, 1961). The timing and choice of which parental X chromosome is inactivated has been studied in mouse development (reviewed in Monk and Grant, 1990). It was shown that X-chromosome inactivation is linked to differentiation, occurring initially as preferential paternal X-inactivation in the trophectoderm, then in the primary endoderm, followed by random X-inactivation in the embryonic cell lineages that give rise to the fetus (Monk and Harper, 1979). Although both X chromosomes are active in females during early preimplantation development, the situation of two active X chromosomes is not compatible with differentiation in development (Shao and Takagi, 1990), nor with long-term culture of female embryonic stem (ES) cells (Rastan and Robertson, 1985). It may also be less than optimal in the early embryo; the effect of a double dosage of X chromosome genes in the mouse zygote has been implicated in the slower development of female embryos in comparison to males (Burgoyne *et al.*, 1995).

In mouse somatic cells, the choice of X chromosome to be inactivated (maternal or paternal) is random, whereas, in the extra-embryonic tissues, the paternally

Table 1. Expression of growth factors in human oocytes and preimplantation embryos (+

Gene		Level of detection	Stage detected
OCT4	Transcription regulator	mRNA	1-cell to 10-cell+
OCT6	Transcription regulator		10-cell+
CD44	Cell surface glycoprotein	Protein	Oocytes, 1-cell to blastocyst
EGF	Epidermal growth factor	mRNA and protein	Oocyte, 8-cell and blastocyst
TGF-α	Transforming growth factor-α		Oocyte, 8-cell and blastocyst
EGF-R	Epidermal growth factor receptor		Oocyte, 8-cell and blastocyst
c-mos	Cell cycle gene	mRNA	Oocytes
cyclin B1	Cell cycle gene		Oocytes to 6-cell
β-actin	House-keeping gene		
c-fms	Colony-stimulating factor 1 receptor		2-cell to blastocyst
SCF	Stem cell factor		2-cell, 8-cell to blastocyst
c-kit	Stem cell factor receptor		2-cell to blastocyst
IL-6 + IL-6R	Interleukin IL6 + receptor	mRNA	Blastocyst only
LIF-R	Leukaemia inhibitory factor receptor component		Blastocyst only
gp130	IL-6R, LIF-R component		Morula to blastocyst
TNF-α	Tumour necrosis factor-α		4-cell to morula
TNF-Rp80	Tumour necrosis factor-α receptor		2- to 4-cell stage
TNF-Rp60	Tumour necrosis factor-α receptor		6- to 8-cell stage
HOXA4	Homeobox genes	mRNA	Oocyte and 4-cell
LIF-Rβ	Leukaemia inhibitory factor receptor components	mRNA	Oocyte, 4-cell and blastocyst
gp130			Oocyte, 4-cell and blastocyst
HLA-G	Non-classical class I major histocompatibility complex molecule	mRNA and protein	Oocytes and blastocyst
IGF-2	Insulin-like growth factor and receptors	mRNA	Oocyte to blastocyst
IGF-1R			Oocyte to blastocyst
IGF-2R			Oocyte to blastocyst
Insulin receptor			8-cell to blastocyst

represents all further stages analysed)

Comments	Reference
	Abdel-Rahman *et al.* (1995)
Speculated to play a role in implantation	Campbell *et al.* (1995)
Expression of EGF, TGF-α and their common receptor EGF-R, suggests autocrine stimulation of preimplantation development	Chia *et al.* (1995)
c-mos expression in oocytes not detected after 6–8 cell indicating loss of maternally inherited mRNA by this stage	Heikiheimo *et al.* (1995)
Expression of c-fms suggests possible paracrine stimulation of preimplantation development by colony stimulating factor, CSF-1, expressed in the oviduct and uterine epithelium. No transcripts of CSF-1 were detected in preimplantation embryos	Sharkey *et al.* (1995)
SCF not expressed in mouse preimplantation embryos. Autocrine stimulation of development in human via c-kit receptor	
Interleukin functions restricted to blastocyst stage. Note conflicting results for LIF-R and gp130 with van Eijk *et al.* (1996)	
TNF-α, inconsistent results, indicates low level of transcripts	
Probably maternally inherited transcripts detected, not zygotic transcripts	Verlinsky *et al.* (1995)
Conflicting results to Sharkey *et al.*, (1995) suggests presence of functional LIF-R in preimplantation embryos which may interact with LIF expression from uterine epithelia	Van Eijk *et al.* (1996)
Expression associated with increased cleavage rate — possibly a functional homologue of mouse Qa-2 antigen	Juriscova *et al.* (1996)
No expression of insulin or IGF-1 detected. Same pattern of expression as in the mouse. Autocrine stimulation of preimplantation development by IGF-2 via its receptor. Possible paracrine stimulation of preimplantation embryos by insulin and IGF-1 released from the uterus	Lighten *et al.* (1997)

human preimplantation development; for example, it is clear from these expression studies that EGF, TGF-α, and SCF may act on their respective receptors, EGF-R (a common receptor for both EGF and TGF-α) and c-kit, to stimulate the development of human preimplantation embryos in an autocrine fashion (Chia et al., 1995; Sharkey et al., 1995). In addition, the expression of c-fms, the receptor for colony-stimulating factor-1 (CSF-1), but not of CSF-1 itself, in human preimplantation embryos suggests that CSF-1 expressed in oviduct and uterine epithelia may exert a paracrine effect on the developing embryo (Sharkey et al., 1995). However, one must be careful when attributing significance to the absence of transcripts of a specific gene since detection depends on the sensitivity of the detection procedures used. Sharkey et al. (1995) conclude that the absence of LIF-R and gp130 (components of the LIF receptor) transcripts early in human preimplantation development, indicates a lack of effect of LIF released from uterine epithelia prior to the blastocyst stage. However, presumably using a more sensitive RT–PCR protocol, van Eijk et al. (1996) detected transcripts for both LIF-Rβ and gp130 in the unfertilized oocyte and 4-cell stage preimplantation embryo, indicating the possible presence of a functional LIF receptor at this stage of development on which LIF may act.

Such conflicting results emphasize the need for experimental controls which reflect the sensitivity of the experimental procedures used. In other words, the absence of transcripts for a specific gene in an early stage preimplantation embryo is not significant if the positive reaction control involves amplification of mRNA from 100 cells. Ideally, the positive controls used should show the presence of cDNA from a single cell from a suitable cell line.

5. cDNA libraries

To overcome the problem of limited availability of preimplantation embryos for the study of gene expression, we have embarked on a project to construct PCR-generated cDNA libraries from single human preimplantation embryos. cDNA libraries have been made from as few as 10 mouse oocytes or preimplantation embryos (Belyavsky et al., 1989; Brady et al., 1990; Revel et al., 1995). So far, we have constructed cDNA libraries for eight oocytes and single 4-cell, 7-cell and blastocyst preimplantation embryos (J. Adjaye, R. Daniels, V. Bolton and M. Monk, unpublished data).

Parental genomic libraries may also be obtained from cumulus cells (maternal) and sperm (paternal) in order to determine the presence of common polymorphisms distinguishing the two parental alleles of a specific gene. In addition, the sex of the single embryo libraries can be determined by analysing the expression of the ZFY and ZFX genes, both of which are expressed as early as the 1-cell stage (Ao et al., 1994).

The cDNA libraries may be used to investigate stage- and tissue-specific expression. During preimplantation development, the two key developmental events are the changes in cell shape at compaction of the 8-cell embryo and the segregation of cells to the inner cell mass (ICM) and the trophectoderm of the blastocyst. The trophectoderm cells are typically epithelial cells and, following proliferation, form extra-embryonic structures such as the placenta. The ICM is initially a non-differentiated, proliferating stem cell population with the capacity to form all the tissues of the adult organism. Subtractive hybridization (Smith and Gridley, 1992) of cDNA from two embryo libraries at different stages of preimplantation development, will allow the isolation of genes expressed at specific stages of preimplantation development. In addition, by making cDNA libraries from isolated ICM and trophectoderm cells, genes expressed in a tissue-specific manner in the ICM and trophectoderm of the blastocyst may also be isolated.

6. Conclusion

Highly sensitive, single-cell procedures must be developed in order to carry out research on preimplantation embryos. The limited studies of gene expression in human preimplantation embryos carried out so far have already revealed differences between humans and other mammalian species and, hence, have underlined the need to study human embryos directly. The results must be interpreted with caution, since the expression of a particular gene in early embryos may not necessarily indicate a specific function at this stage; there is evidence suggesting a temporary derepression of gene activity, perhaps due to the absence of regulatory mechanisms at the onset of development. Studies on the pattern of expression of the *XIST* gene in human embryos show differences from the mouse which may yield additional information on the regulation of early X-inactivation. In addition to this more academic research, expression studies with clinical relevance have been initiated concerning growth-promoting interactions between the developing preimplantation embryo and the cells lining the oviduct and the uterus. These studies will lead to improved culture conditions, thereby increasing the efficiency of IVF treatment of infertility and preimplantation diagnosis of genetic defects. Finally, the creation of cDNA libraries from individual human preimplantation embryos will provide a limitless and readily available resource for studies on gene expression during preimplantation development and obviate the need for fresh embryonic material in the future.

Acknowledgements

We thank all our colleagues in the Molecular Embryology Unit at the Institute of Child Health for stimulating discussions and for their support and collaboration in our research.

References

Abdel Rahman B, Fiddler M, Rappolee D, Pergament E. (1995) Expression of transcription regulating genes in human preimplantation embryos. *Mol. Hum. Reprod.* **10**: 2787–2792.

Ao A, Erickson RP, Winston RML, Handyside AH. (1994) Transcription of paternal Y-linked genes in the human zygote as early as the pronucleate stage. *Zygote* **2**: 281–287.

Belyavsky A, Vinogradova T, Rajewsky K. (1989) PCR-based cDNA library construction: general cDNA libraries at the level of a few cells. *Nucleic Acids Res*. **17**: 2919–2932.

Borsani G, Tonlorenzi R, Simmler MC *et al.* (1991) Characterization of a murine gene expressed from the inactive X chromosome. *Nature* **351**: 325–329.

Brady G, Barbara M, Iscove N. (1990) Representative in vitro cDNA amplification from individual hematopoietic cells and colonies. *Methods Mol. Cell. Biol*. **2**: 17–25.

Braude PR, Bolton V, Moore S. (1988) Human gene expression first occurs between the four- and eight-cell stages of preimplantation development. *Nature* **332**: 459–461.

Braude PR, Monk M, Pickering SJ, Cant A, Johnson MH. (1989) Measurement of HPRT activity in the human unfertilized oocyte and pre-embryo. *Prenatal Diagn*. **9**: 839–850.

Brockdorff N, Ashworth A, Kay GF *et al.* (1991) Conservation of position and exclusive expression of mouse *Xist* from the inactive X chromosome. *Nature* **351**: 329–331.

Brook JD, McCurrah ME, Harley HG *et al.* (1992) Molecular basis of myotonic dystrophy: expansion of a trinucleotide (CTG) repeat at the 3′ end of a transcript encoding a protein kinase family member. *Cell* **68**: 799–808.

Brown CJ, Willard HF. (1994) The human X inactivation centre is not required for maintenance of X chromosome inactivation. *Nature* **368**: 154–156.

Brown CJ, Ballabio A, Rupert JL, Lafreniere RG, Grompe M, Tonlorenzi R, Willard HF. (1991) A gene from the region of the X-inactivation centre is expressed exclusively from the inactive X chromosome. *Nature* **349**: 38–44.

Brown CJ, Hendrich BD, Rupert JL, Lafreniere RG, Xing Y, Lawrence JB, Willard HF. (1992) The human XIST gene: analysis of a 17kb inactive X-specific RNA that contains conserved repeats and is localised within the nucleus. *Cell* **71**: 527–542.

Burgoyne PS, Thornhill AR, Kalnus Boudreau S, Darling SM, Evans EP. (1995) The genetic basis of XX–XY differences present before gonadal sex differentiation in the mouse. *Philos. Trans. R. Soc. Lond. B*. **250**: 253–261.

Buzin CH, Mann JR, Singer-Sam J. (1994) Quantitative RT–PCR assays show Xist RNA levels are low in mouse female adult tissue, embryos and embryoid bodies. *Development* **120**: 3529–3536.

Campbell S, Swann HR, Aplin JD, Seif MW, Kimber SJ, Elstein M. (1995) CD44 is expressed throughout pre-implantation human embryo development. *Hum. Reprod*. **10**: 425–430.

Chia CM, Winston RML, Handyside AH. (1995) EGF, TGF-α and EGFR expression in human preimplantation embryos. *Development* **121**: 299–307.

Clemson CM, McNeil JA, Willard HF, Lawrence JB. (1996) *XIST* RNA paints the inactive X chromosome at interphase: evidence for a novel RNA involved in nuclear/chromosome structure. *J. Cell Biol*. **132**: 259–275.

Clepet C, Schafer AJ, Sinclair AH, Palmer MS, Lovell-Badge R, Goodfellow P. (1993) The human SRY transcript. *Hum. Molec. Genet*. **2**: 2007–2012.

Daniels R, Kinis T, Serhal P, Monk M. (1995) Expression of the myotonin protein kinase gene in preimplantation embryos. *Hum. Mol. Genet*. **4**: 389–393.

Daniels R, Lowell S, Bolton V, Monk M. (1997a) Transcription of tissue-specific genes in human preimplantation development. *Hum. Reprod*. (in press).

Daniels R, Zuccotti M, Kinis T, Serhal P, Monk M. (1997b) *XIST* expression in human oocytes and preimplantation embryos. *Am. J. Hum. Genet*. **61**: 33–39.

Dardik A, Schultz RM. (1991) Blastocoel expansion in the preimplantation mouse embryo: stimulatory effect of TGF-alpha and EGF. *Development* **113**: 919–930.

Davidson RG, Nitowsky HM, Childs B. (1963) Demonstration of two populations of cells in the human female heterozygous for glucose-6-phosphatase dehydrogenase variants. *Proc. Natl Acad. Sci. USA* **50**: 481–485.

Fiddler M, Abdel-Rahman B, Rappolee D, Pergament E. (1995) Expression of SRY transcripts in preimplantation human embryos. *Am. J. Med. Genet*. **55**: 80–84.

Fry RC, Batt PA, Fairclough RJ, Parr RA. (1992) Human leukemia inhibitory factor improves the viability of cultured ovine embryos. *Biol. Reprod*. **46**: 470–474.

Fu Y, Friedman DL, Richards S *et al.* (1993) Decreased expression of myotonin-protein kinase messenger RNA and protein in adult form of myotonic dystrophy. *Science* **260**: 235–238.

Goto T, Wright E, Monk M. (1997) Paternal X chromosome inactivation in human trophoblastic cells. *Mol. Hum. Reprod*. **3**: 77–80.

Harper MI, Fosten M, Monk M. (1982) Preferential paternal X inactivation in extra-embryonic tissues of early mouse embryos. *J. Embryol. Exp. Morphol.* **67**: 127–138.

Harrison KB. (1989) X-chromosome inactivation in the human cytotrophoblast. *Cytogenet. Cell Genet.* **52**: 37–41.

Harvey MB, Kaye PL. (1991) Mouse blastocysts respond metabolically to short-term stimulation by insulin and IGF-1 through the insulin receptor. *Mol. Reprod. Dev.* **29**: 253–258.

Harvey MB, Leco KJ, Arcellana-Panlilio MY, Zhang X, Edwards DR, Schultz GA. (1995) Roles of growth factors during peri-implantation development. *Mol. Hum. Reprod.* **10**: 712–718.

Heikiheimo O, Lanzendorf SE, Baka SG, Gibbons WE. (1995) Cell cycle genes *c-mos* and cyclin-B1 are expressed in a specific pattern in human oocytes and preimplantation embryos. *Mol. Hum. Reprod.* **10**: 699–707.

Juriscova A, Casper RF, Maclusky NJ, Mills GB, Librach C. (1996) HLA-G expression during preimplantation human embryo development. *Proc. Natl Acad. Sci. USA* **93**: 161–165.

Kauma SW, Matt DW. (1995) Coculture cells that express leukemia inhibitory factor (LIF) enhance mouse blastocyst development *in vitro. J. Assist. Reprod. Genet.* **12**: 153–156.

Kay GF, Penny GD, Patel D, Ashworth A, Brockdorff N, Rastan S. (1993) Expression of Xist during mouse development suggests a role in the initiation of X chromosome inactivation. *Cell* **72**: 171–182.

Kay GF, Barton SC, Surani MA, Rastan S. (1994) Imprinting and X chromosome counting mechanisms determine Xist expression in early mouse development. *Cell* **77**: 639–650.

Kidder GM. (1992) The genetic program for preimplantation development. *Devel. Genet.* **13**: 319–325.

Koopman P, Ashworth A, Lovell-Badge R. (1991) The ZFY gene family in humans and mice. *Trends Genet.* **7**: 132–136.

Larson RC, Ignotz GG, Currie WB. (1992) Platelet derived growth factor (PDGF) stimulates development of bovine embryos during the fourth cell cycle. *Development* **115**: 821–826.

Lighten AD, Hardy K, Winston RML, Moore GE. (1997) Expression of mRNA for the insulin-like growth factors and their receptors in human preimplantation embryos. *Mol. Reprod. Devel.* **47**: 134–139.

Lyon MF. (1961) Gene action in the X chromosome of the mouse (*Mus musculus* L.). *Nature* **190**: 372–373.

Monk M. (1986) Methylation and the X chromosome. *BioEssays* **4**: 204–208.

Monk M, Grant M. (1990) Preferential X-chromosome inactivation, DNA methylation and imprinting. *Development* (Suppl:) 55–62.

Monk M, Harper MI. (1979) Sequential X chromosome inactivation coupled with cellular differentiation in early mouse embryos. *Nature* **281**: 311–313.

Monk M, Handyside A, Hardy K, Whittingham D. (1987) Preimplantation diagnosis of deficiencies of hypoxanthine phosphoribosyl transferase in a mouse model for Lesch–Nyhan syndrome. *Lancet* **2**: 423–425.

Nothias JY, Majumder S, Kaneko KJ, DePamphilis ML. (1995) Regulation of gene expression at the beginning of mammalian development. *J. Biol. Chem.* **270**: 22077–22080.

Paria BC, Dey SK. (1990) Preimplantation embryo development in vitro: cooperative interactions among embryos and role of growth factors. *Proc. Natl Acad. Sci. USA* **87**: 4756–4760.

Penny GD, Kay GF, Sheardown SA, Rastan S, Brockdorff N. (1996) Requirement for Xist in X chromosome inactivation. *Nature* **379**: 131–137.

Rastan S, Robertson EJ. (1985) X-chromosome deletions in embryo-derived (EK) cell lines associated with lack of X-chromosome inactivation. *J. Embryol. Morphol.* **90**: 379–388.

Revel F, Renard J-P, Duranthon V. (1995) PCR-generated cDNA libraries from reduced numbers of mouse oocytes. *Zygote* **3**: 241–250.

Ropers HH, Wolff G, Hitzeroth HW. (1978) Preferential X inactivation in human placenta membranes: is the paternal X inactive in early embryonic development of female mammals? *Hum. Genet.* **43**: 265–273.

Sabouri LA, Mahadevan MS, Narang M, Lee DSC, Surh LC, Korneluk RG. (1993) Effect of the myotonic dystrophy (DM) mutation on mRNA levels of the DM gene. *Nature Genet.* **4**: 233–238.

Sakkas D, Trounson AO. (1990) Co-culture of mouse embryos with oviduct and uterine cells prepared from mice at different days of pseudopregnancy. *J. Reprod. Fertil.* **90**: 109–118.

Salvatori S, Biral D, Furlan S, Marin O. (1994) Identification and localisation of the myotonic dystrophy gene product in skeletal and cardiac muscles. *Biochem. Biophys. Res. Commun.* 203: 1365–1370.

Sermon K, Lissens W, Tarlatzis B, Braude PR, Devroy P, Van Steirteghem A, Liebaers I. (1992) β-N-acetylhexosaminodase activity in human oocytes and preimplantation embryos. *Hum. Reprod.* 7: 1278–1280.

Shao C, Takagi N. (1990) An extra maternally derived X chromosome is deleterious to early mouse development. *Development* 110: 969–975.

Sharkey AM, Dellow K, Blayney M, Macnamee M, Charnock-Jones S, Smith SK. (1995) Stage-specific expression of cytokine and receptor messenger ribonucleic acids in human preimplantation embryos. *Biol. Reprod.* 53: 955–962.

Smith DE, Gridley T. (1992) Differential screening of a PCR-generated mouse embryo cDNA library: glucose transporters are differentially expressed in early postimplantation mouse embryos. *Development* 116: 555–561.

Sprinks MT, Sellens MH, Dealtry GB, Fernandez N. (1993) Preimplantation mouse embryos express *Mhc* class I genes before the first cleavage division. *Immunogenetics* 38: 35–40.

Takagi N, Sasaki M. (1975) Preferential inactivation of the paternally derived X chromosome in the extraembryonic membranes of the mouse. *Nature* 256: 640–642.

Telford NA, Watson AJ, Schultz GA. (1990) Transition from maternal to embryonic control in early mammalian development: a comparison of several species. *Mol. Reprod. Dev.* 26: 90–100.

Tesarik J, Kopecny V, Blachot M, Mandelbaum J. (1986) Activation of nucleolar and extra-nucleolar RNA synthesis and changes in ribosomal content of human embryos developing in vitro. *J. Reprod. Fertil.* 78: 463–470.

van Eijk MJT, Mandelbaum J, Salat-Baroux J, Belaisch-Allart J, Plachot M, Junca AM, Mummery CL. (1996) Expression of leukaemia inhibitory factor receptor subunits LIFRβ and gp130 in human oocytes and preimplantation embryos. *Mol. Hum. Reprod.* 2: 355–360.

Verlinsky Y, Morozov G, Gindilis V *et al.* (1995) Homeobox gene expression in human oocytes and preembryos. *Mol. Reprod. Dev.* 41: 127–132.

West JD, Freis WI, Chapman VM, Papaioannou VE. (1977) Preferential expression of the maternally derived X chromosome in the mouse yolk sac. *Cell* 12: 873–882.

West JD, Flochart JH, Angell RR, Hillier SG, Thatcher SS, Glasier AF, Rodger MW, Baird DT. (1989) Glucose phosphate isomerase activity in mouse and human eggs and pre-embryos. *Hum. Reprod.* 4: 82–85.

Whiting EJ, Waring JD, Tamai K, Somerville MJ, Hincke M, Staines WA, Ikeda JE, Korneluk RG. (1995) Characterization of myotonic dystrophy kinase (DMK) protein in human and rodent muscle and central nervous tissue. *Hum. Mol. Genet.* 4: 1063–1072.

Wood SA, Kaye PL. (1989) Effects of epidermal growth factor on preimplantation mouse embryos. *J. Reprod. Fertil.* 85: 575–582.

Zhang XY, Loflin PT, Gehrke CW, Andrews PA, Ehrlich M. (1987) Hypermethylation of human DNA sequences in embryonal carcinoma cells and somatic tissues but not in sperm. *Nucleic Acids Res.* 15: 9429–9449.

10

The expression of *Hox* genes in post-implantation human embryos

Edoardo Boncinelli and Peter Thorogood

1. Introduction

During embryonic development, a vast diversity of cell types is produced from a relatively homogeneous pool of precursor cells. All events of differentiation and patterning at the level of single cells or groups of cells are preceded by developmental decisions that involve entire regions (or 'morphogenetic fields') of the forming organism. In recent years, the cellular and molecular mechanisms by which specific types of tissues and organs are generated and organized have been explored and a number of principles and developmental pathways of general validity have been discovered. Interestingly, some of these principles appear to be evolutionarily conserved.

A key role in these developmental processes is played by the so-called regulatory genes; genes acting through the control of the expression of other genes lying hierarchically downstream from them and sometimes termed 'target' genes. Regulatory genes generally code for transcription factors, that is nuclear proteins able to recognize specific DNA sequences, bind to them and modulate, through this specific binding, the level of expression of the corresponding target genes. A number of regulatory genes are homeobox genes. Homeobox genes are regulatory genes characterized by the presence of a specific, evolutionarily conserved, DNA sequence termed the 'homeobox', able to code for a protein domain of some 60 amino acid residues, termed the 'homeodomain'. It is through the action of their homeodomain that the protein products of the homeobox genes, the homeoproteins, bind to the regulatory regions of specific genes and control their expression.

Decades of patient and inspired genetic analysis have led to the identification of a number of developmental regulatory genes necessary for the establishment of the anterior–posterior axis and for its patterning in a succession of appropriate

Molecular Genetics of Early Human Development, edited by T. Strachan, S. Lindsay and D.I. Wilson.
© 1997 BIOS Scientific Publishers Ltd, Oxford.

anatomical regions or body segments (Lewis, 1978). The first homeobox genes were reported in 1984 in the fruit fly, *Drosophila melanogaster* (McGinnis *et al.*, 1984). According to the timing of their action and to their function, these genes may be assigned to three general categories: maternal genes, segmentation genes and homeotic genes (Akam, 1987; Ingham and Martinez Arias, 1992; McGinnis and Krumlauf, 1992; St Johnston and Nüsslein-Volhard, 1992). In *Drosophila*, the function of maternal genes, already active in the body of the mother, is to introduce into the unfertilized egg all the molecules necessary and sufficient to specify the anterior and posterior extremities, as well as the dorso–ventral polarity, of the future organism. After fertilization, segmentation genes control the subdivision of the fly embryo into a succession of potential body segments, where the number of segments and their relative order must be correct. If all of these genes have operated in the appropriate manner, the genes of the third category, the homeotic genes, finally provide the morphogenetic specification or developmental identity of the various segments. Through the action of homeotic genes each and every segment along the body of the fly acquires its full endowment of anatomical structures, including appropriate appendages, and, as a consequence, acquires its full and specific functionality. Mutations in these genes cause the transformation of a given body segment into a more or less perfect copy of a different body segment. Thus, body parts inappropriate for their particular location within the body plan develop; the various dysmorphic phenotypes are collectively referred to as 'homeotic mutants'. This unique characteristic of homeotic genes has attracted the attention of biologists for decades and they have been among the first genes to be isolated and molecularly characterized as the technology of the so-called 'genetic engineering' became available.

Eight of these homeotic genes, namely *labial, proboscipedia, Deformed, Sex combs reduced, Antennapedia, Ultrabithorax, abdominal-A* and *abdominal-B*, are located near each other in two contiguous chromosomal loci collectively called *HOM-C*, for homeotic complex (Akam, 1987, 1989). All eight *HOM* genes have been shown to contain a homeobox and, as a consequence, to be homeobox genes. Many more homeobox genes have been subsequently identified in flies and in all multicellular organisms, including vertebrates and mammals (Duboule, 1994a). It is now obvious that homeobox genes simply represent a subset of a wider class of the regulatory genes, that is genes coding for transcription factors. For this reason it is even more remarkable that all eight *HOM* genes are in fact homeobox genes. We do not yet have an explanation for that or for a number of unique features exhibited by these genes and by their vertebrate homologues, the *Hox* genes. In this chapter we focus on one particular aspect of *Hox* gene expression during human embryonic development, but initially we provide a context from animal studies for this discussion.

2. The *Hox* genes: background information

The *Hox* genes stand out among the numerous vertebrate homeobox genes as the true homologues of the *Drosophila HOM* homeotic genes. In mouse and man there are 39 *Hox* genes evenly distributed in four different chromosomal loci termed

Hox loci or clusters: *Hoxa, Hoxb, Hoxc* and *Hoxd* in the mouse, *HOXA, HOXB, HOXC* and *HOXD* in man. *HOXA*, contains 11 genes, *HOXB* contains 10 genes, whereas both the *HOXC* and *HOXD* clusters each contain nine genes. In the human, the four clusters map to chromosomes 7 (7p14), 17 (17q21), 12 (12q31) and 2 (2q31), respectively (Acampora *et al.*, 1989; Boncinelli *et al.*, 1991; Zeltser *et al.*, 1996). Within each cluster the various genes are organized in the same transcriptional orientation from their 5' to their 3' extremities. Originally, the *Hox* genes had been identified by the primary sequence of their homeodomains. In fact, most *Hox* genes encode proteins containing a homeodomain closely related to the archetypal *Antennapedia* homeodomain. However, since the resemblance is relatively slight for *Hox* genes located at the ends of the four clusters, it is better to define *Hox* genes both on the basis of the primary sequence of the encoded products and of their physical location on chromosomes (Boncinelli *et al.*, 1991).

The four clusters are highly homologous and can be easily aligned. Corresponding genes in different *Hox* clusters share the highest sequence similarity and are described as 'paralogues' and belonging to the same 'paralogous group'. There are 13 such paralogous groups, numbered from 1 to 13 starting from the 3' end of the various clusters. Only paralogous groups 4, 9 and 13 contain the complete set of four members across the four clusters, while in the remaining paralogous groups one or more members are missing in living mammals (*Figure 1*).

The homology of the four *Hox* clusters, and the very existence of the paralogous groups, strongly suggest that the four clusters arose through duplication of a single ancestral cluster. Further support for this conclusion comes from the finding that the non-*Hox* genes adjacent to *Hox* clusters have also been duplicated during evolution along with the clusters with which they are syntenic (Ruddle *et al.*, 1994). The duplicated clusters were subsequently distributed separately on four different chromosomes. Most interestingly, the four *Hox* clusters are also homologous to the *HOM-C* complex of insects (*Figure 1*). In fact, most of the vertebrate *Hox* genes can be easily shown to be the homologues of some homeotic genes of *Drosophila*; for example, *Hox* genes of the paralogous group 1 are homologues of the fly homeotic gene *labial*, whereas *Hox* genes of the paralogous group 4 are homologues of the fly gene *Deformed*. Less clear is the exact correspondence of *Hox* genes of groups 5–8 with homeotic genes, whereas *Hox* genes belonging to groups 9–13 appear to be all related to the homeotic gene *Abdominal-B*, even if with a varying degree of similarity. On the basis of sequence data, evolutionary arguments and expression data, the *Hox* gene family may be subdivided into three subfamilies: one constituted by the genes belonging to groups 1–4, a second constituted by genes of groups 5–8 and a third comprising the remaining genes of groups 9–13 (Boncinelli *et al.*, 1991).

In evolutionary terms, all of this probably means that the vertebrate *Hox* clusters and the fly *HOM-C* complex derive from an ancestral locus, containing at least four different *Hox* genes, and which was already in existence prior to the divergence of the evolutionary lineages leading to insects on one hand and to vertebrates on the other. In fact, a homologous *HOM/Hox* cluster containing a reduced number of homeobox genes has been identified in the nematode *Caenorhabditis elegans* (Kenyon, 1994). The evolutionary implications of such a conserved genomic organization appear to be even more intriguing, as we have

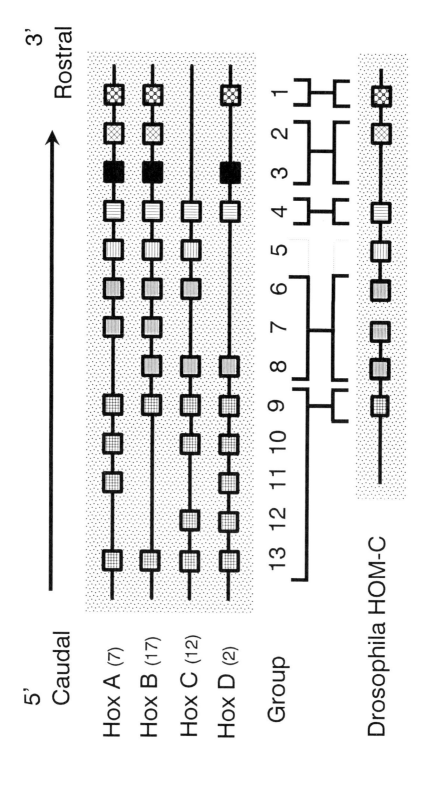

Figure 1. See facing page for legend.

identified two human homologues of the fly *even skipped* homeobox segmentation gene at the 5' end of the *HOXA* and *HOXD* loci, albeit in the opposite orientation to the *Hox* genes. These genes, termed *EVX1* and *EVX2*, respectively, are, in turn, homologous to *Evx 1* and *Evx 2* identified in the mouse (Bastian and Gruss, 1990; D'Esposito *et al.*, 1991; Faiella *et al.*, 1991).

We can conclude that these genes remained clustered in a restricted chromosomal region and approximately in the same order for hundreds of millions of years. This is the first of the unique features exhibited by these genes. Even more striking is a second feature, termed 'colinearity'. The *HOM-C* genes in *Drosophila* and the *Hox* genes in vertebrates are expressed in relatively extended expression domains along the major body axis of the corresponding embryos. These gene-specific expression domains, which in vertebrates include discrete regions of the neural tube, are characterized by sharp anterior boundaries and a progressively decreasing expression in posterior regions. Within the *HOM-C* complex, the homeotic genes of *Drosophila* are aligned in a physical 3' to 5' order that is co-linear with the anterio–posterior order of the body segments whose morphogenetic fate they control. This colinearity is manifest within the anterio–posterior sequence of expression domains along the axis at the time that morphogenetic fate of these individual segments is specified. Similarly, within each vertebrate *Hox* cluster, genes are arranged in a 3' to 5' order that is colinear with the anterior limits or 'cut-offs' of the expression domains along the embryonic axis and with the relative anterio–posterior position of that part of the body plan whose morphogenetic fate they apparently control (Akam, 1989; Boncinelli *et al.*, 1988; Duboule and Dollé, 1989; Graham *et al.*, 1989). Furthermore, the members of each paralogous group display equivalent expression domains and may therefore have some common functions during development (possibly reflecting the common origin of each group from a single ancestral gene) resulting in the possibility of 'functional redundancy' (Hunt and Krumlauf, 1992), as revealed by transgenesis experiments in the mouse.

In addition to this striking structural colinearity, the homeobox genes of the various vertebrate *Hox* loci also exhibit a second colinear phenomenon, termed 'temporal colinearity'. Thus, the first *Hox* genes to be turned on during embryonic development are those located at the 3' end of the four clusters, which are expressed in anterior regions. Subsequently the *Hox* genes lying in progressively more 5' locations within the clusters are turned on progressively later in development, each paralogous group exhibiting a progressively more posterior expression domain (Boncinelli *et al.*, 1991; Dekker *et al.*, 1992; Duboule, 1994a; Izpisua-

---

Figure 1. Schematic representation of the human *HOX* gene clusters (*A*, *B*, *C*, and *D*) and their chromosomal locations (in brackets). Clusters are aligned so that paralogous genes belonging to the same group are aligned vertically. The *Drosophila HOM-C* complex, thought to resemble the single ancestral complex, and the known homologies with vertebrate genes, is shown below. The direction of transcription is indicated by the arrow. Genes situated at the 3' end of the clusters are expressed more rostrally and earlier than genes at the 5' end. It is the genes of paralogous groups 1–4 which have anterior cut-offs within the branchial region and which are thought to have a critical role in morphogenetic specification during craniofacial development. (Diagram supplied by Paul Hunt.)

Belmonte *et al.*, 1991; Simeone *et al.*, 1990). Elucidation of precisely how temporal and structural colinearity of *Hox* gene expression is achieved and controlled during the development of any particular morphogenetic field remains a formidable challenge.

Hox genes have also been shown to be activated in a colinear way upon treatment with retinoic acid, a metabolite of vitamin A, both in cultured cells and *in vivo* (Boncinelli *et al.*, 1991; Simeone *et al.*, 1990, 1991). Thus 3′ (anteriorly expressed) *Hox* genes of the four clusters are activated by this drug more promptly and more efficiently than 5′ (posteriorly expressed) genes. The effects of retinoids are mediated by two families of nuclear receptors, the retinoic acid receptors (RAR) and the retinoid X receptors (RXR). The retinoic acid/receptor complexes operate as transcription factors which operate by binding to retinoic acid response elements (RARE) in the promoter region of the target gene. Significantly, a number of *Hox* genes have now been demonstrated to contain such RAREs (Marshall *et al.*, 1994; Morrison *et al.*, 1996; Ogura and Evans, 1995) and retinoids are well known to have pleiotropic effects on development and differentiation. However, despite an impressive body of evidence for an activating effect of added exogenous retinoic acid on the expression of *Hox* genes in mouse and frog embryos (e.g. Conlon and Rossant, 1992; Durston *et al.*, 1989; Marshall *et al.*, 1992; Studer *et al.*, 1994), it is still a matter of debate whether endogenous retinoids play a role in the sequential activation of these genes in *normal* embryogenesis (see various reviews in Jones, 1997). Nevertheless, it has been argued that retinoic acid signalling and *Hox* gene responses are an ancient and conserved mechanism establishing the anterio–posterior body axis in chordates generally (reviewed in Shimeld, 1996).

The 3′-anterior / 5′-posterior rule of colinearity between gene position and the sequence of anterior cut-offs means that *Hox* gene expression is manifested in nested domains along the embryonic axis and within individual morphogenetic fields, such as the vertebral column, limb and genital tubercle (Dollé *et al.*, 1991; Duboule, 1992; Kessel and Gruss, 1990). The differing degrees of overlap of expression domains of the various paralogous groups within any individual field suggests that the role of *Hox* genes in morphogenetic specification is fulfilled by a combinatorial mode of action (Hunt and Krumlauf, 1992), whereby the particular combination of (*Hox*) genes expressed in any particular region determines the fate of the cells at that position, presumably in the context of the cells' lineage and developmental history. The nature of the causal link between *Hox* codes and the actual morphogenesis and patterns of differentiation specified remains largely unknown. There is still widespread ignorance over the identity of downstream target genes, although evidence is emerging that other regulatory genes, including other *Hox* genes (e.g. Zappavigna *et al.*, 1991), as well as structural genes, such as those encoding cell adhesion molecules (e.g. Jones *et al.*, 1992) may be targets. More recently, the interesting suggestion has been made that *Hox* genes may control morphogenesis by primarily determining local patterns of growth (Duboule, 1994b).

The greater part of the work on *Hox* gene organization, expression and function has been carried out using the mouse because of the opportunities it provides for transgenesis; consequently it is from this species that we have the most complete understanding. Nevertheless the rather fragmentary information that we have from the other model systems used, principally the chick, *Xenopus* and the zebrafish, is broadly in accord with that from the mouse and this is taken to reflect a degree of functional conservation (reviewed by Holland and Garcia-Fernandez, 1996). The organ systems which have been studied are principally the vertebral column, limb, genital tubercle, gut and branchial region of the head (reviewed in Dietrich and Kessel, 1997; Ferretti and Tickle, 1997; Thorogood, 1997). Whilst coverage of most of these is outside the scope of this chapter since they have not been studied in the human, the murine vertebral column does provide us with an exemplar of a *Hox* code in action (Burke *et al.*, 1995). The colinear expression of the *Hox* genes of all four clusters along the anterio–posterior/rostro–caudal axis provides a linear sequence of cut-offs in the paraxial somitic tissue. The cataloguing of these cut-offs at the emerging sclerotomal boundaries has been used as the basis of a combinatorial '*Hox* code' for the developing mouse vertebral column (Kessel and Gruss, 1990). This model has been tested by both transgenic (e.g. Lufkin *et al.*, 1992) and teratogenic strategies (using exogenous retinoic acid to change the *Hox* code; Kessel and Gruss, 1990, 1991) and, in most instances, the resultant changes of the code elicit a homeotic transformation of vertebral phenotype. Generally, the phenotypic consequences of transgenic inactivation coincide with the rostral expression boundary, suggesting that this is the critical region of the expression domain.

Within the other principal axial tissue, the neural tube, the anterior cut-offs for any individual *Hox* gene and for all members of its paralogous group, are usually more rostral than in the adjacent paraxial mesoderm or its sclerotomal derivative. Of particular interest to us here is the branchial region of the developing head where *Hox* genes of the first four paralogous groups are expressed according to a specific pattern (reviewed in Krumlauf, 1994; Lumsden and Krumlauf, 1996). This region also constitutes the *only* morphogenetic field within the human embryo for which there has been a comprehensive study of *Hox* gene expression (see later); for that reason, it is the first system in which we can begin to assess the role of *Hox* genes in human development (and dysmorphogenesis?) with any degree of knowledge about involvement of this family of regulatory genes.

(Given the flexure of the embryonic axis at cervical and cranial levels that takes place during development, the terms 'anterior' and 'posterior' become imprecise with regard to axis and in the ensuing part of the chapter, on craniofacial development, they are replaced with the terms 'rostral' and 'caudal' respectively)

3. *Hox* gene expression in the branchial region

The head of the vertebrate embryo is a modular structure with different cell lineages giving rise to a variety of tissues and structures. We can rationalize it, somewhat simplistically, into the anterior neural tube, which develops into the (fore-, mid- and hind-) brain, the associated mesoderm, which is the primary source of myogenic cells giving rise to the facial musculature, and the neural crest, emanating from the apices of the neural folds and giving rise to virtually all of the connective tissue of the head including the dermis and most of the skull. The neural crest can be viewed as the major player in the patterning of craniofacial tissues other than the CNS. A variety of experimental strategies indicate that its own morphogenetic specification is established very early on, even prior to the emigration of crest cells from the neural folds and whilst they are still within the neural primordium (reviewed in Thorogood, 1997).

For the purposes of analysis, the embryonic head can be divided into the 'rostral head' which goes on to form the upper face, and the 'branchial region', which gives rise to the middle and lower face and neck structures. The former develops at the axial level of the forebrain and associated mesoderm and does not use members of the *Hox* gene family in its development although other homeobox genes, principally of the *Otx* and *Emx* families, are actively involved in morphogenetic specification in this region (reviewed by Bally-Cuif and Boncinelli, 1997). In contrast, the branchial region, comprising the midbrain and hindbrain boundary and the hindbrain itself, together with the branchial arches (derived from ancestral gill arches in a chordate ancestor) *does* use *Hox* genes to specify fate in its constituent neurectoderm-derived cell populations. Both hindbrain and branchial arches display a metamerism or segment-like organization and it is the relationship between the two, and the derivation of the hindbrain or rhombencephalic crest, which are pivotal to our understanding of the development of this region.

Development of the hindbrain, or rhombencephalon, is marked by the appearance of a series of periodic constrictions and bulges that subdivide the region into distinct areas, termed 'rhombomeres' and numbered r1 to r8 from anterior to posterior. Clonal analysis in chick embryos has shown that as soon as each rhombomere boundary forms, cell movement across it is restricted. Thus, the hindbrain soon becomes subdivided into at least five compartments corresponding to rhombomeres r2–r6 (Fraser *et al.*, 1990). Subsequently, neuronal development occurs in a pattern that correlates with this segmentation; each of the branchial motor nerves arises from adjacent pairs of rhombomeres, the V nerve from r2 and r3, the VII nerve from r4 and r5 and the IX nerve from r6 and r7 (Lumsden and Keynes, 1989). Clues regarding the cellular basis of the formation of hindbrain segments have come from grafting experiments. These data suggest that some alteration in cellular properties, for example cell adhesion, might underlie the restriction of cell movement across rhombomere boundaries (Guthrie and Lumsden, 1991). The rhombencephalon appears to be the only primarily segmented region of the developing neural tube of vertebrates and it might, therefore, be predicted to possess several characteristics which are phylogenetically highly conserved.

In the embryonic mouse head, the genes of the first four paralogous groups of *Hox* genes have been found to display cut-offs within the hindbrain at the interfaces between the rhombomeres (Wilkinson *et al.*, 1989). *Hox* genes of the paralogous group 5 are expressed in anterior regions of the embryonic spinal cord with an anterior boundary coinciding with the boundary between hindbrain and future spinal cord (Hunt *et al.*, 1991a,b; Wilkinson *et al.*, 1989). *Hox* genes of group 4 are expressed in r7 and r8 with an anterior boundary coinciding with the boundary between r7 and r6. *Hox* genes of group 3 are expressed in r6 and r5 with an anterior boundary coinciding with the boundary between r5 and r4. *Hox* genes of group 2 are generally expressed in r4 and r3 with an anterior boundary coinciding with the boundary between r3 and r2, with *HoxB2* also extending to r2.

Hox genes of group 1 constitute an exception to the rule and are expressed mainly in r4 (*Figure 2*).

Significantly, neural crest cells derived from the hindbrain migrate into branchial arches at equivalent axial levels and give rise to much of the facial mesenchyme (*Figure 2*). Thus, crest cells from rhombomeres 1 and 2 migrate to colonize arch 1, the mandibular arch, expressing no *Hox* genes [although other homeobox genes such as those of the *Dlx* family, may have a role here (Qui *et al.*, 1995)]. More caudally, the migrating crest cells maintain the combination of *Hox* gene expression which characterized their rhombomere of origin. Cells from rhombomeres 4 and 6 migrate into arches two and three respectively (Lumsden *et al.*, 1991). A segmental organization of crest migration is ensured by crest cells from rhombomeres 3 and 5 either being lost through apoptosis (Graham *et al.*, 1993; Lumsden *et al.*, 1991) or merging with the streams of cells emigrating from rhombomeres 4 and 6 (Sechrist *et al.*, 1993). Events caudal to arch three are speculative but the model predicts that cells from rhombomeres 7 and 8 migrate into the fourth arch and possibly the sixth arch (modern convention holds that the fifth arch was lost during evolution, although there is no hard evidence for this).

The branchial *Hox* code defined by the patterns of combinatorial *Hox* gene expression (Hunt *et al.*, 1991a) has been interpreted as a developmental strategy whereby positional specification made axially within the neural tube, and manifest as the code, is transmitted to the periphery (the arches) via the migrating neural crest, thus ensuring anatomical registration during development of the branchial region (Noden, 1992; *Figure 2*). It is therefore seen as an integral part of the mechanisms whereby a vertebrate embryo builds a head and face (Hunt *et al.*, 1991b) and is assumed to have been established early in chordate phylogeny. Even though downstream targets are largely unknown, various transgenic (e.g. Gendron-Maguire *et al.*, 1993; Le Mouellic *et al.*, 1992) and experimental embryological (e.g. Hunt *et al.*, 1995) strategies indicate that this interpretation is largely correct. To date, all of this is based on experimental analysis and manipulation of various animal model systems, chiefly the mouse. The assumption has been made, in the absence of any comprehensive, comparative data, that the paradigms emerging are applicable to all vertebrates including humans. However, given the pivotal role proposed for these genes, and for the combinatorial manner in which they are expressed, it can be argued that elucidation of the human *Hox* gene expression pattern is an essential prerequisite to understanding the genetic control of human facial development and the dysmorphogenesis of this complex process. Furthermore, such data will also facilitate an assessment of the degree to which the mouse system can justifiably provide a model for human craniofacial development.

4. The human branchial *Hox* code

In one of the first multigene expression studies to be carried out using human embryonic tissue, we analysed the human branchial arch *Hox* code at 4 and 5 weeks postfertilization, using specific probes against message from eight *HOX* genes whose mouse homologues have been previously identified as having cut-offs at the

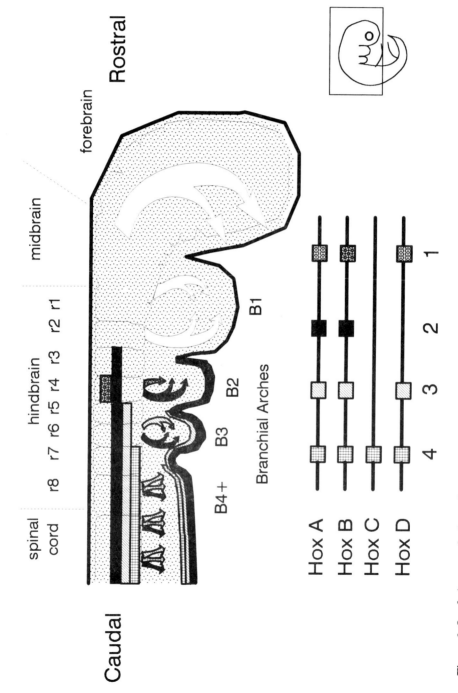

Figure 2. See facing page for legend.

relevant axial levels. These genes include not only representatives of each of the first four paralogous groups (1–4) but also two of those groups (2 and 3) in their entirety (Vieille-Grosjean *et al.*, 1997).

In the human embryo at 4 weeks postfertilization the branchial arches display their maximum development before fusion and remodelling (Thorogood, 1997). Using *in situ* hybridization on serial sections through the hindbrain and through the branchial arches of human embryos at this stage, we defined the expression pattern of *HOXB1*, *HOXA2*, *B2*, *HOXA3*, *B3*, *D3*, *HOXB4* and *C4* at this stage. *Figure 3* indicates the coronal plane of section used for most embryos studied and typical autoradiographic *in situ* results for a single gene, *HOXA3*, are shown in *Figure 4*.

The expression of one of the three genes constituting paralogous group 1, *HOXB1*, was studied and found to be expressed as a single stripe in rhombomere 4 in the hindbrain. No expression was detected in branchial arches one, two and three, and expression was only detected in the caudal ectoderm of the fourth branchial arch, and in the ectoderm lining the fourth cleft and covering the sixth arch. Expression was also detected in the dorsal endoderm of the foregut at the same axial level.

Both paralogues of group 2 were studied. Within the hindbrain, expression of *HOXA2* displayed a clear rostral cut-off at the boundary between rhombomeres 1 and 2, whereas the cut-off for *HOXB2* occurred at the boundary between rhombomeres 2 and 3. In fact, *HOXB2* expression was higher in rhombomeres 3, 4 and 5 than in more caudal rhombomeres. In contrast to these expression differences within the hindbrain, in the branchial arches the two paralogues had equivalent patterns of expression, with signal being detected in arches two, three and four.

Paralogous group 3 contains three paralogues, all of which were examined. *HOXA3*, *B3* and *D3* displayed the same rostral limit of expression in the hindbrain, between rhombomeres 4 and 5 but with differences in the intensity of expression (*Figure 4*). Thus, expression of *HOXA3* was higher in rhombomeres 5 and 6 than in the more caudal segments, *HOXB3* was expressed less intensely in rhombomeres 5 and 6, and expression of *HOXD3* was only minimally detectable

---

Figure 2. The branchial *Hox* code as elucidated in the mouse embryo, illustrating the relationship between expression domains of the 3'-most *Hox* genes in the rhombomeres (r1–r8) in the developing hindbrain and branchial arches (B1–B4+), and their position within their respective clusters (labelled A–D) on individual chromosomes. Paralogous groups are defined vertically across the clusters and these 3'-most paralogous groups are numbered 1–4. Note the correspondence (as indicated by hatching) between an individual gene in its cluster, its paralogues in other clusters, and their spatial expression domain in the rhombomere, in streams of migrating crest cells (arrows) and in the ectodermal covering of each arch (migration of crest cells from particular axial levels is indicated by arrows). Spatial colinearity is seen in the rostral-to-caudal sequence of anterior cut-offs of expression (from right to left) and the location of the individual genes 3' to 5' along each chromosome (also from right to left). Note that no *Hox* genes are expressed in those crest cells migrating from the first two rhombomeres into the first arch (which is thought to reflect a 'default' branchial arch specification) or in those migrating rostrally from the mid- and forebrain. The location of the diagram *vis-a-vis* the rest of the embryo is indicated in the inset. (Diagram supplied by Paul Hunt.)

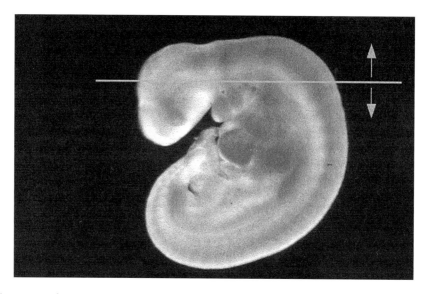

Figure 3. Left lateral view of a 4-week human embryo showing maximum development of the branchial arches, of which the first and second can be seen to be particularly prominent. The horizontal blue line indicates the coronal plane of sectioning used to obtain the images shown in *Figure 4*.

at this stage in rhombomere 5. All three genes were expressed in the branchial arches with a rostral limit of expression between arch two and arch three but again displayed regional differences in intensity; for example, *A3* was expressed uniformly throughout the arches two, three and four (*Figure 4*), but the level of expression of *HOXB3* and *D3* was lower in the third than in the fourth arch.

Finally, the expression pattern of two of the four genes belonging to paralogous group 4 was assessed. Both *HOXB4* and *C4* were found to be expressed in the more caudal rhombomeres. The expression pattern of *HOXB4* showed a clear cut-off at the boundary between rhombomere 6 and 7, whereas *HOXC4* displayed no equivalent cut-off, with its most rostral and weak expression observed in rhombomere 8 and gradually increasing caudally along the neural tube. At the level of the branchial arches, both genes were expressed in domains caudal to arch four.

Figure 4. Autoradiographic *in situ* hybridization revealing expression of *HOXA3* in the branchial region of a 4-week postconception human embryo (dark-field microscopy). (a) Coronal section through the hindbrain showing the sharp anterior cut-off of gene expression within the neuroepithelial wall of one side of the rhombencephalon, at the junction between rhombomeres 4 and 5 (r4/5) as indicated by the arrowhead. The otocyst (OV) lateral to the hindbrain and ganglion IX/X provide landmarks; rostral is to the left. (b) Similar plane of section, but at a more ventral level, showing expression in the branchial arches on one side of the oropharynx of the same embryo; again, rostral is to the left. Note the dramatic anterior cut-off (arrows) at the interface between arches 2 and 3 (b2/b3). (Modified from Vieille-Grosjean *et al.*, 1997.)

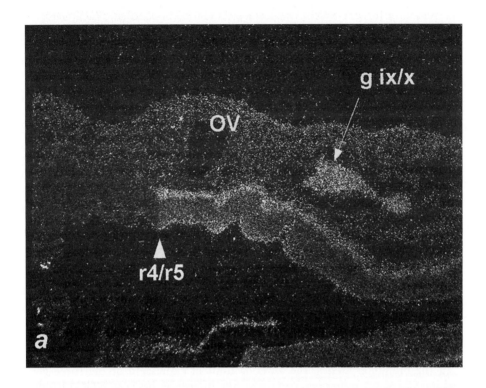

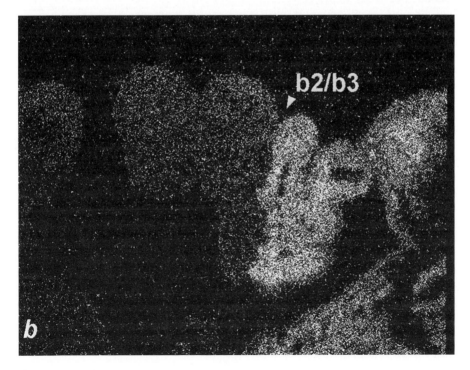

HOXB4 was expressed in a large stripe of mesenchyme surrounding the pharyngeal region of foregut and in the surface ectoderm; in contrast, *HOXC4* expression was only detected in the dorsal endoderm of the foregut and surface ectoderm.

These data on expression of genes from paralogous groups 1–4, for the hindbrain and for the branchial arches, constitute a map (*Figure 5*) of the human branchial *Hox* code based on the eight *Hox* genes studied to date (Vieille-Grosjean *et al.*, 1997)

Critical events in branchial arch morphogenesis commence during the fifth week of development in human embryos. These involve allometric growth, fusion and remodelling of the arches as the first stages in face and neck development ensue (Thorogood, 1997). Accordingly, the pattern of expression of selected *HOX* genes was also assessed during this later stage. In particular, we wished to explore the possibility that members of an individual paralogous group might become differentially expressed with time. Therefore, expression of all three paralogues of group 3, *HOXA3*, *B3* and *D3*, was assessed.

We found that within the hindbrain, expression of all three genes was identical to that defined previously in 4-week-old embryos, and with a rostral limit of expression at rhombomere 5. Expression of *HOXD3* in rhombomere 5 was more obvious at this later stage of development, but nevertheless remains less intense than the signal in rhombomere 6. More ventral sections, including those through the branchial arches, revealed that expression of all three genes in the more caudal neural tube was high, with expression also being observed in dorsal root ganglia.

In the branchial arches, the anterior limit of expression of the three genes remained between arches two and three, but differences between *HOXA3* and *D3* had become more apparent with further development. A clear cut-off was observed for both *HOXA3* and *HOXB3* between arches two and three, whereas expression of *HOXD3* was only found in the caudal mesenchyme of the third arch, displayed a poorly defined cut-off and was absent in all other components of this arch. Similarly, in the fourth arch, *HOXD3* was strongly expressed in the mesenchyme but, again, was not detectable in the other tissues of the arch.

Both the thymus and parathyroid glands are derived from the third branchial pouch (Larsen, 1993) and *Hoxa3* has been implicated in the normal development of both these derivatives in the mouse (Chisaka and Capecchi, 1991). Accordingly, we analysed the expression of these genes in the different components of the branchial arches at this axial level, namely the surface ectoderm, the ectoderm-lined external clefts, the mesenchyme and the endoderm-lined branchial pouches. Significantly, there was differential expression within the third branchial pouch, with both *HOXA3* and *B3* being expressed by the endodermal epithelium but *HOXD3* not expressed. More caudally, within the tracheo-laryngeal groove (the site where the tracheal primordium exits from the primitive gut as a diverticulum) and the mesodermal arytenoid swellings (which will subsequently form the arytenoid cartilages), *HOXB3* alone was weakly expressed, with no signal found for *A3* or *D3*. In the thyroid primordium, our limited data revealed that *HOXD3* was definitely not expressed, but limited tissue availability meant that we were not able to obtain data on *A3* or *B3* expression within this structure.

Rhombomeres

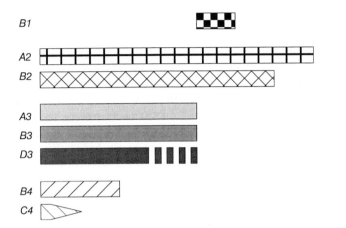

Branchial arches

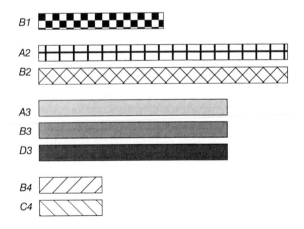

Figure 5. The *Hox* code for the human branchial region, summarizing expression data for eight genes which are co-expressed in this region at 4 weeks of embryonic development. The nested nature of the expression domains, and the common domains shared by members of any individual paralogous group (2, 3 or 4), are clearly evident in this diagram. (Modified from Vieille-Grosjean *et al.*, 1997.)

5. Evaluation of human branchial *Hox* gene expression

There has been a dramatic increase in our understanding of the molecular and cellular mechanisms underlying craniofacial development and most of this has been achieved within a relatively short period of time. However, the relevance of much of this new knowledge to the clinical dysmorphologist remains unclear. Initial excitement over the possibility that *Hox* gene mutations might underlie a number of disease states and dysmorphic syndromes has simply not been vindicated (Redline *et al.*, 1992). Ironically, it has been the *PAX* gene family which has proved to be much more fertile with regard to identification of genes implicated in human birth defects (see Strachan and Read, 1994), although, thus far published expression patterns of PAX genes in postimplantation human embryos are limited to a preliminary report by Gerard *et al.* (1995). It is striking that *HOX* genes have not been implicated directly in any human craniofacial syndrome, although secondary, downstream involvement cannot be excluded. Yet, as the recent discovery of *HOXD13* as the gene for human synpolydactyly (Muragaki *et al.*, 1996) demonstrates, there is still much to be elucidated about *HOX* genes and human development.

Indeed, a number of significant observations about early human craniofacial development emerge from this work. Perhaps the single, most important conclusion to be drawn is that the basic architecture of *HOX* gene expression is essentially the same in the branchial region of the 4-week human embryo as it is for other vertebrate species at equivalent stages of development (Hunt *et al.*, 1991a,b; Wilkinson *et al.*, 1989). From the genes studied here, this conclusion applies equally to both the rhombomeric region of the hindbrain and the branchial arches which will go on to form much of the face and neck. Clearly the vertebrate branchial '*Hox* code', as it has been elucidated from work in the mouse (Hunt *et al.*, 1991a,b; Wilkinson *et al.*, 1989) and from the incomplete data from avian (Hunt *et al.*, 1995) and amphibian species (Godsave *et al.*, 1994), remains highly conserved in the spatial pattern of expression domains and in the timing of its expression. Although the existing data can tell us little about conservation of function, the strong correlations described here with the published mouse data, suggest that the repertoire of *HOX* gene expression serves an equivalent function in the embryonic assembly of the head and face in the human and the mouse. In other words, whatever it is that makes the human and the mouse faces different, it lies downstream of the branchial *Hox* code. Attention now needs to be focused on the identity of these downstream target genes and analysis of their deployment during development.

Throughout, the issue of functional homology remains critical, because, for all the current interest in homology, there has previously been no thorough appraisal of genetic conservation on this scale in humans. Although, given the apparently fundamental role of the branchial *Hox* code, conservation might have been expected, a rigorous evaluation of gene expression was necessary in order to assess that likelihood. This study provided, for the first time and in any system, the confirmation of such an expectation for human development. However, a single demonstration such as this cannot be sufficient justification to make generalized assumptions about other genes with homologues in animal model systems.

Certainly we cannot assume *a priori* that conclusions reached here will hold true for other regulatory genes implicated directly or indirectly in human birth defects. Should species-specific differences in expression of any particular genes be found, it will necessitate a re-evaluation of the assumptions we make about the validity of the mouse model for that particular organ system or morphogenetic field.

One less obvious, but nevertheless valuable, outcome is that these data allow us to assess the extent to which the mouse can justifiably be used as a model system to understand human craniofacial development and its dysmorphogenesis. On the basis of the similarities in gene expression described here it would seem to be reasonable to assume that models of normal development based on *Hox* gene function, emerging from mouse work where experimental analyses can be carried out, can safely be extrapolated to the human. This will include data and paradigms emerging from transgenic manipulations but should not necessarily be extended uncritically to all transgenic lines being adopted as models for human craniofacial anomalies. Indeed, apart from one notable exception (Chisaka and Capecchi, 1991; Manley and Capecchi, 1995; and see later) there is little evidence to date of a human craniofacial syndrome with a phenotype matching that produced by any mouse *Hox* knockout, suggesting that loss-of-function mutations in *HOX* genes are unlikely to be a common explanation for dysmorphic syndromes affecting the head and face.

The model of a combinatorial code of *Hox* gene expression underlying the morphogenetic specification of the branchial arches, and therefore underpinning much of early facial development, is based largely on the orderly migration of crest cells from individual rhombomeres to specific arches as reported from animal studies (e.g. Lumsden *et al.*, 1991).

Ironically, these human data allow us to follow up several predictions of that model and to extend it significantly. In the first instance, although data exist for the first three branchial arches, migration of crest cells from rhombomeres 7 and 8 to the fourth arch is ill-defined and largely a prediction based on assumptions about serial homology (Hunt *et al.*, 1991a,b). This qualification applies also to the assumed expression of group 4 *Hox* genes in the fourth arch (Hunt *et al.*, 1991a). However, the analysis reported here extends what is known about the branchial arch *Hox* code by assessing gene expression in the fourth arch. Surprisingly, group 4 genes are not expressed in the fourth arch and we could only find rostral cut-offs for *HOXB4* and *C4* immediately caudal to arch 4. Although it appears from this that assumptions about serial homology of the branchial arches break down at the level of the more caudal arches, this may not necessarily be the case. Down-regulation of expression of group 4 genes might occur during migration to, or shortly following, their arrival in this arch destination, but the existing data do not allow us to distinguish between these various possibilities; the expression of *HOXB1* in rhombomere 4 but with a rostral cut-off only detectable in arch four not arch two, as might otherwise be predicted, may be explicable in the same way. Clearly, further data are needed on the *timing* of gene expression in the human system. Nevertheless, it is interesting that whereas the anterior boundaries of expression domains of *HOX* genes belonging to any individual paralogous group coincide in branchial arches, they do not in the neuroepithelium of the rhombencephalon itself (*Figure 5*) and the significance of this difference remains unclear at this stage.

A second advance emerging from the data is the observation that within a single paralogous group (in this instance, group 3) the various paralogues become differentially expressed as development proceeds. Thus, within days of the entire group being expressed within the branchial arch region and sharing an expression domain, we find modulation of expression domains emerging; for example, the *HOXD3* gene is only expressed in the caudal region of the third arch unlike its two paralogues which are expressed throughout. Perhaps more significantly, however, *HOXA3* and *D3* are differentially expressed within the lining of the third branchial pouch. This is significant, not only in terms of what it reveals about normal development. It has fundamental implications for how we interpret dysmorphogenesis, not just in transgenic mouse phenotypes but also for how we define the aetiology of human craniofacial problems. The transgenic inactivation of *Hoxa3* in the mouse produces a dysmorphic phenotype which includes thymic aplasia, thyroid hypoplasia and hypoparathyroidism and it has been proposed as a model for the DiGeorge syndrome in humans (Chisaka and Capecchi, 1991). Interestingly, the *Hoxd-3* knock-out displays a different phenotype in which these tissues are unaffected and the primary defects are in the cervical vertebrae (Condie and Capecchi, 1993). Given that there are major endodermal contributions to both thymus and the parathyroid glands from the third branchial pouch (Larsen, 1993), our finding that of the two human paralogues, only *HOXA3* is expressed within the epithelial lining of this pouch, is significant for two reasons. First, it explains why individual inactivation of each of the two paralogues in the mouse produces two completely different phenotypes with regard to branchial pouch derivatives (Condie and Capecchi, 1994). Furthermore, our data, showing that *HOXD3* is definitely not expressed in the thyroid primordium, explain why inactivation of this paralogue in the mouse leaves thyroid development unaffected. Second, the data provides unique evidence supporting the contention that inactivation of *Hoxa3* is a valuable model for DiGeorge (Chisaka and Capecchi, 1991; Manley and Capecchi, 1995), since here we show that it is indeed expressed in the pouch from which the human thymus and parathyroids are derived. *HOXA3* is not the DiGeorge gene but nevertheless may serve to regulate unidentified DiGeorge genes (at 22q11 and 10p13), which is perhaps why the *Hoxa3* knockout displays some of the features of the syndrome. Should this be the case, then the noted variation in syndromic phenotype may reflect subtle polymorphisms in a regulatory *Hox* gene which would not necessarily be apparent when using the mouse model.

As stated earlier, *HOX* genes have not been widely implicated as common loci for mutations generating human craniofacial abnormalities. Nevertheless, given the undoubted importance of this family of genes within the genetic hierarchy of craniofacial morphogenesis, delineation of those genes comprising the human branchial *Hox* code, and knowledge of their spatiotemporal patterns of expression, must be an essential and integral part of understanding the molecular events underlying craniofacial dysmorphogenesis in humans. How such information might be used can be illustrated by reference to retinoic acid-induced abnormalities (Morriss-Kay, 1993). Certain retinoids are among the most potent craniofacial teratogens known and a characteristic retinoid-induced dysmorphic phenotype is clinically recognized in the human (Lammer *et al.*, 1985). Given that retinoic acid response elements are contained within the enhancer of some *Hox* genes, and that *in vitro* activation of *Hox* genes in a 3′ to 5′ sequence can be induced by increasing concentrations of retinoic acid (discussed earlier), then a more insightful explanation of how elevated retinoid levels operate as craniofacial teratogens becomes possible (Brickell and Thorogood, 1997). Moreover, the identification of mutation in *HOXD13* causing expansion of a polyalanine stretch in the encoded homeoprotein (Muragaki *et al.*, 1996), suggests that members of the *HOX* gene family may still be implicated in other birth defects.

Finally, it should be pointed out that the study illustrates that, even within the constraints imposed by the rarity and finite availability of well-preserved embryonic material, judicious use can generate comprehensive and valuable data. The original research paper (Vieille-Grosjean *et al.*, 1997) constituted one of the first published reports of a multigene expression analysis of human embryonic or fetal tissue at any stage of development, with tissue from any individual embryo being used to define the expression of up to six genes. This illustrates the value of careful coordinated use of such rare material and demonstrates the very considerable potential for the use of banked human embryonic tissues in the long term. The bank from which this material was obtained is a facility funded to provide tissue for the research community at large, and currently over thirty projects from around Europe and the UK are registered. We anticipate that this level of usage will continue to increase as more dysmorphology genes are identified and researchers endeavour to understand the developmental mechanisms relating altered genotype to disrupted phenotype. Analysis of *HOX* gene expression within other morphogenetic fields of the human embryo, especially the limb, will undoubtedly be a part of this development.

Acknowledgements

The authors research described here was supported by the European Union, the UK Medical Research Council and the Programme Telethon-Italia. PT would like gratefully to acknowledge Paul Hunt, Isabelle Vieille-Grosjean and Joe Chan for providing illustrative material.

References

Acampora D, D'Esposito M, Faiella A *et al.* (1989) The human HOX gene family. *Nucleic Acids Res.* **17**: 10385–10402.

Akam M. (1987) The molecular basis for metameric pattern in the *Drosophila* embryo. *Development* **101**: 1–22.

Akam M. (1989) Hox and HOM: homologous gene clusters in insects and vertebrates. *Cell* **57**: 347–349.

Bally-Cuif L, Boncinelli E. (1997) Transcription factors and head formation in vertebrates. *BioEssays* **19**: 127–135.

Bastian H, Gruss P. (1990) A murine even-skipped homologue, Evx1, is expressed during early embryogenesis and neurogenesis in a biphasic manner. *EMBO J.* **9**: 1839–1852.

Boncinelli E, Somma R, Acampora D, Pannese M, D'Esposito M, Faiella A, Simeone A. (1988) Organization of human homeobox genes. *Hum. Reprod.* **3**: 880–886.

Boncinelli E, Simeone A, Acampora D, Mavilio F. (1991) HOX gene activation by retinoic acid. *Trends Genet.* **7**: 329–334.

Brickell P, Thorogood P. (1997) Retinoic acid and retinoic acid receptors in craniofacial development. In: *Retinoic Acid and Retinoic Acid Receptors in Development* (ed. E Jones). *Seminars in Cellular and Developmental Biology*, Vol. 8. Academic Press, New York (in press).

Burke AC, Nelson CE, Morgan BA, Tabin C. (1995) *Hox* genes and the evolution of vertebrate axial morphology. *Development* **121**: 333–346.

Chisaka O, Capecchi M. (1991) Regionally restricted *developmental* defects resulting from targetted disruption of the mouse homeobox gene Hox1.5. *Nature* **350**: 473–479.

Condie BG, Capecchi MR. (1993). Mice homozygous for a targeted disruption of Hoxd-3 (Hox-4.1) exhibit anterior transformations of the first and second cervical vertebrae, the atlas and the axis. *Development*, **119**: 579–595.

Condie BG, Capecchi MR. (1994). Mice with targeted disruptions in the paralogous genes hoxa-3 and hoxd-3 reveal synergistic interactions. *Nature* **370**: 304–307.

Conlon RA, Rossant J. (1992) Exogenous retinoic acid rapidly induces anterior ectopic expression of murine Hox-2 genes in vivo. *Development* **116**: 357–368.

Dekker EJ, Pannese M, Houtzager E, Timmermans A, Boncinelli E, Durston A. (1992) Xenopus Hox-2 genes are expressed sequentially after the onset of gastrulation and are differentially inducible by retinoic acid. *Development* **1992** (Suppl.) 195–202.

D'Esposito M, Morelli F, Acampora D, Migliaccio E, Simeone A, Boncinelli E. (1991) EVX2, a human homeobox gene homologous to the even-skipped segmentation gene, is localized at the 5′ end of HOX4 locus on chromosome 2. *Genomics* **10**: 43–50.

Dietrich S, Kessel M. (1997) The vertebral column. In: *Embryos, Genes and Birth Defects* (ed. P Thorogood). John Wiley & Sons, Chichester, pp. 281–302.

Dollé P, Izpisúa-Belmonte J-C, Tickle C, Brown J, Duboule D. (1991). Hox-4 genes and the morphogenesis of mammalian genitalia. *Genes Dev.* **5**: 1767–1776.

Duboule D. (1992). The vertebrate limb: a model system to study the Hox/HOM gene network during development and evolution. *BioEssays* **14**: 375–384.

Duboule D. (ed.) (1994a) *Guidebook to the Homeobox Genes*. Oxford University Press, Oxford.

Duboule D. (1994b) Temporal colinearity and the phylotypic progression; a basis for the stability of a vertebrate Bauplan and the evolution of morphologies through heterochrony. *Development* (Suppl.): 135–142.

Duboule D, Dollé P. (1989). The structural and functional organization of the murine Hox gene family resembles that of *Drosophila* homeotic genes. *EMBO J.* **8**: 1497–1505.

Durston AJ, Timmermans JPM, Hage WJ, Hendriks HFJ, de Vries NJ, Heideveld M, Niewkoop PD. (1989) Retinoic acid causes an anterior posterior transformation in the developing nervous system. *Nature* **340**: 140–144.

Faiella A, D'Esposito F, Rambaldi M *et al.* (1991) Isolation and mapping of EVX1, a human homeobox gene homologous to even-skipped, localized at the 5′ end of HOX1 locus on chromosome 7. *Nucleic Acids Res.* **19**: 6541–6545.

Ferretti P, Tickle C. (1997) The limbs. In: *Embryos, Genes and Birth Defects* (ed. P Thorogood), John Wiley & Sons, Chichester, pp. 101–132.

Fraser S, Keynes R, Lumsden A. (1990) Segmentation in the chick embryo hindbrain is defined by cell lineage restrictions. *Nature* **344**: 431–435.

Gendron-Maguire M, Mallo M, Zhang M, Gridley T. (1993). Hoxa-2 mutant mice exhibit homeotic transformation of skeletal elements derived from cranial neural crest. *Cell* **75**: 1317–1331.

Gerard M, Abitbol M, Delezoide A-L, Dufier J-L, Mallet J, Vekemans M. (1995) PAX genes expression during human embryonic development a preliminary report. *CR Acad. Sci. Paris, Sciences de la vie* **318**: 57–66.

Godsave S, Dekker EJ, Holling T, Pannese M, Boncinelli E, Durston A. (1994). Expression patterns of Hoxb genes in the *Xenopus* embryo suggest roles in anteroposterior specification of the hindbrain and in dorsoventral patterning of the mesoderm. *Dev. Biol.* **166**: 465–476.

Graham A, Papalopulu N, Krumlauf R. (1989) The murine and *Drosophila* homeobox gene complexes have common features of organization and expression. *Cell.* **57**: 367–378.

Graham A, Heyman I, Lumsden A. (1993). Even numbered rhombomeres control the apoptotic elimination of neural crest cells from odd numbered rhombomeres in the chick hindbrain. *Development* **119**: 233–245.

Guthrie S, Lumsden A. (1991) Formation and regeneration of rhombomere boundaries in the developing chick hindbrain. *Development* **112**: 221–229.

Holland PWH, Garcia-Fernandez J. (1996) *Hox* genes and chordate evolution. *Dev. Biol.* **173**: 382–395.

Hunt P, Krumlauf R. (1992). Hox codes and positional specification in vertebrate embryonic axes. *Annu. Rev. Cell. Biol.* **8**: 227–256.

Hunt P, Gulisano M, Cook M, Sham M-H, Faiella A, Wilkinson D, Boncinelli E, Krumlauf R. (1991a) A distinct *Hox* code for the branchial region of the vertebrate head. *Nature* **353**: 861–864.

Hunt P, Whiting J, Nonchev S et al. (1991b) The branchial code and its implications for gene regulation, patterning of the nervous system and head evolution. *Development* (Suppl. 2): 63–77.

Hunt P, Ferretti P, Krumlauf R, Thorogood P. (1995). Restoration of normal Hox code and branchial arch morphogenesis after extensive deletion of hindbrain neural crest. *Dev. Biol.* **168**: 584–597.

Ingham PW, Martinez Arias A. (1992) Boundaries and fields in early embryos. *Cell* **68**: 221–235.

Izpisua-Belmonte JC, Tickle C, Dollé P, Wolpert L, Duboule D. (1991) Expression of homeobox Hox-4 genes and the specification of position in chick wing development. *Nature* **350**: 585–589.

Jones E. (ed.) (1997) *Retinoic Acid and Retinoic Acid Receptors in Development. Seminars in Cellular and Developmental Biology*, Vol. 8. Academic Press, New York (in press).

Jones FS, Predinger EA, Bittner DA, DeRobertis EM, Edelman GH. (1992) Cell adhesion molecules as targets for *Hox* genes: neural cell adhesion molecule promoter activity is modulated by co-transfection with *Hox-2.5* and *-2.4*. *Proc. Natl Acad. Sci. USA* **89**: 2086–2090.

Kenyon C. (1994) If birds can fly, why can't we?: homeotic genes and evolution. *Cell* **78**: 175–180.

Kessel M, Gruss, P. (1990) Murine developmental control genes. *Science* **249**: 374–379.

Kessel M, Gruss P. (1991) Homeotic trasformations of murine prevertebrae and concomitant alteration of Hox codes induced by retinoic acid. *Cell* **67**: 89–104.

Krumlauf R. (1994) *Hox* genes in vertebrate development. *Cell* **78**: 191–201.

Lammer EJ, Chen DT, Hoar RM et al. (1985) Retinoic acid embryopathy. *N. Engl. J. Med.* **313**: 832–841.

Larsen WJ. (1993). *Human Embryology*. Churchill Livingstone, New York, NY.

Le Mouellic H, Lallemand Y, Brulet P. (1992) Homeosis in the mouse induced by a null mutation in the Hox-3.1 gene. *Cell* **69**: 251–264.

Lewis EB. (1978) A gene complex controlling segmentation in *Drosophila*. *Nature* **276**: 565–570.

Lufkin T, Manuel M, Hart CP, Dollé P, LeMeur M, Chambon P. (1992). Homeotic transformation of the occipital bones of the skull by ectopic expression of a homeobox gene. *Nature* **359**: 835–841.

Lumsden A, Keynes R. (1989) Segmental patterns of neuronal development in the chick hindbrain. *Nature* **337**: 424–428.

Lumsden A, Krumlauf R. (1996) Patterning the vertebrate neuraxis. *Science* **274**: 1109–1115.

Lumsden A, Sprawson N, Graham A. (1991). Segmental origin and migration of neural crest cells in the hindbrain region of the chick embryo. *Development* **113**: 1281–1291.

Manley NR, Capecchi MR. (1995). The role of Hoxa-3 in mouse thymus and thyroid development. *Development* **121**: 1989–2003.

Marshall H, Nonchev S, Sham M-H, Muchamore I, Lumsden A, Krumlauf R. (1992) Retinoic acid alters hindbrain *Hox* code and induces transformation of rhombomeres 2/3 into a 4/5 identity. *Nature* **360**: 737–741.

Marshall H, Studer M, Pöpperl H, Aparicio S, Kuroiwa A, Brenner S, Krumlauf R. (1994). A conserved retinoic acid response element required for early expression of the homeobox gene Hoxb-1. *Nature* **370**: 567–571.

McGinnis W, Krumlauf R. (1992) Homeobox genes and axial patterning. *Cell* **68**: 283–302.

McGinnis W, Levine MS, Hafen E, Kuroiwa A, Gehring WJ. (1984) A conserved DNA sequence in homeotic genes of the *Drosophila* Antennapedia and Bithorax complexes. *Nature* **308**: 428–433.

Morrison A, Moroni MC, Ariza-McNaughton L, Krumlauf R. (1996) In vitro and transgenic analysis of a human *HOXD4* retinoid-responsive enhancer. *Development* **122**: 1895–1907.

Morriss-Kay GM. (1993). Retinoic acid and craniofacial development. *BioEssays* **15**: 9–15.

Muragaki Y, Mundlos S, Upton J, Olsen BR. (1996) Altered growth and branching patterns in synpolydactyly caused by mutations in HOXD13. *Science* **272**: 548–551.

Noden DM. (1992) Spatial integration among cells forming the cranial peripheral nervous system. *J. Neurobiol.* **24:** 248–261.

Ogura T, Evans RM. (1995). Evidence for two distinct retinoic acid response pathways for HOXB1 gene regulation. *Proc. Natl Acad. Sci. USA* **92:** 392–396.

Qiu M, Bulfone A, Martinez S, Meneses JJ, Shimamura K, Pedersen RA, Rubenstein JLR. (1995) Null mutation of *Dlx-2* results in abnormal morphogenesis of proximal first and second branchial arch derivatives and abnormal differentiation in the forebrain. *Genes Dev.* **9:** 2523–2538.

Redline RW, Neish A, Holmes L, Collins T. (1992) Biology of disease: Homeobox genes and congenital malformations. *Lab. Invest.* **66:** 659–670.

Ruddle FH, Bentley KL, Murtha MT, Risch N. (1994) Gene loss and gain in the evolution of vertebrates. *Development* (Suppl.): 155–161.

Sechrist J, Serbedzija GN, Scherson T, Fraser S, Bronner-Fraser, M. (1993). Segmental migration of the hindbrain neural crest does not arise from segmental generation. *Development* **118:** 691–703.

Shimeld S. (1996) Retinoic acid, HOX genes and the anterior-posterior axis in chordates. *BioEssays* **18:** 613–616.

Simeone A, Acampora D, Arcioni L, Andrews PW, Boncinelli E, Mavilio F. (1990) Sequential activation of HOX2 homeobox genes by retinoic acid in human embryonal carcinoma cells. *Nature* **346:** 763–766.

Simeone A, Acampora D, Nigro V, Faiella A, D'Esposito M, Stornaiuolo A, Mavilio F, Boncinelli E. (1991) Differential regulation by retinoic acid of the homeobox genes of the four *Hox* loci in human embryonal carcinoma cells. *Mech. Dev.* **33:** 215–228.

St Johnston D, Nüsslein-Volhard C. (1992) The origin of pattern and polarity in the *Drosophila* embryo. *Cell* **68:** 201–220.

Strachan T, Read AP. (1994) PAX genes. *Curr. Opin. Genet. Dev.* **4:** 427–438.

Studer M, Pöpperl H, Marshall H, Kuriowa A, Krumlauf R. (1994) Role of a conserved retinoic acid response element in rhombomere restriction of *Hoxb-1*. *Science* **256:** 1728–1732.

Thorogood P. (1997) The head and face. In: *Embryos, Genes and Birth Defects* (ed. P Thorogood). John Wiley & Sons, Chichester, pp. 197–230.

Vieille-Grosjean I, Hunt P, Gulisano M, Boncinelli E, Thorogood P. (1997) Branchial HOX gene expression and human craniofacial development. *Dev. Biol.* **183:** 49–60.

Wilkinson DG, Bhatt S, Cook M, Boncinelli E, Krumlauf R. (1989) Segmental expression of *Hox-2* homoeobox-containing genes in the developing mouse hindbrain. *Nature* **341:** 405–409.

Zappavigna V, Renucci A, Ispizua-Belmonte J, Urier G, Peschle C, Duboule D. (1991) *Hox-4* genes encode transcription factors with potential auto- and cross-regulatory capacities. *EMBO J.* **10:** 4177–4188.

Zeltser L, Desplan C, Heintz N. (1996) *Hoxb-13*: a new Hox gene in a distant region of the HOXB cluster maintains colinearity. *Development* **122:** 2475–2484.

11

Expression of *Wnt* genes in postimplantation human embryos

Susan Lindsay, Philip Bullen, Majlinda Lako, Julia Rankin, Steve C. Robson and Tom Strachan

1. Introduction

As discussed in Chapter 2, there is good reason to believe that gene expression patterns of orthologous genes will often not be identical in human and mouse and that for developmental control genes, as well as genes underlying developmental disorders, there may be important consequences for the understanding of specific gene function in the two species. As well as providing model systems for studying human genetic disorders, mice (and rats) are increasingly being studied in order to elucidate normal developmental pathways. It is, therefore, of great importance to define the similarities and differences in gene function between human and mouse in order to assess, as accurately as possible, where the mouse model is a true representative of the processes underlying human development and disease. Where this is not the case, there may be profound implications for our understanding and future potential treatments of specific human disorders. Defining and comparing gene expression patterns during human and mouse development is a first step in gathering information about functional differences.

We have a particular interest in studying gene expression in the developing brain. The anatomy of this organ, particularly the forebrain, is very different in human and mouse. By a very early stage of human development (Carnegie Stage 20, approximately 51 embryonic days) the differentiation of the human forebrain is unmistakably different from that of other mammalian embryos and is characterized by a considerable enlargement of the cerebral hemispheres (Marin-Padilla, 1988). This suggests that developmental processes, which lead to the comparatively enormous cerebral capacity of humans, begin very early in development. In addition to the very different capacities of the brains of human and mouse, there are parallel species differences in the morphology and spatial relationships between specific brain regions. Thus, it is particularly important to

---

Molecular Genetics of Early Human Development, edited by T. Strachan, S. Lindsay and D.I. Wilson.
© 1997 BIOS Scientific Publishers Ltd, Oxford.

identify the similarities and differences in expression patterns in mouse and human for genes which have a role in brain development.

Members of the *Wnt* gene family are known to have important roles in brain development in mice and other vertebrates (Parr *et al.*, 1993; and see Moon *et al.*, 1997a and references therein) and we have chosen to investigate to what extent *Wnt* gene expression in the developing human brain parallels that in the mouse. The name '*Wnt*' comes from an amalgamation of *wg* (*Drosophila wingless* gene; Baker *et al.*, 1987; Rijsewijk *et al.*, 1987) and *int* (integration site of mouse mammary tumour virus; Van Ooyen and Nusse, 1984), which very appropriately reflects the roles that members of the *Wnt* gene family have in both normal development and tumorigenesis. *Wnt* genes encode secreted signalling molecules which, it is thought, usually act over a short range. Considerable progress has been made in elucidating the *Wnt* pathway, including the recent identification of proteins which act as Wnt receptors (Bhanot *et al.*, 1996). These are members of a family of genes, *frizzled* (*fz*), and work is in progress to determine whether there are other receptor proteins *in vivo* and also the relationships between individual Wnt proteins and specific members of the fz family. Very recently, a *Xenopus* protein, Frzb, has been identified which is closely related to the extracellular domain of Fz proteins but which has no transmembrane domain. Frzb has been shown to be a secreted protein which, in *Xenopus*, binds to and antagonizes the activity of one member of the Wnt family, XWnt8 (reviewed in Moon *et al.*, 1997b). It remains to be seen whether Frzb can interact with other Wnt proteins and whether there are other *Frzb* genes whose products may interact with specific Wnt proteins. Similar genes have also been found in mouse and human, but in these species they have been named *Fritz* (*mfiz* and *hfiz*, respectively). Mfiz has been shown to inhibit XWnt8 activity in *Xenopus* embryos (Mayr *et al.*, 1997). It is also becoming clear that the original model of a linear *Wnt* pathway is oversimplified and that Wnt proteins are involved in at least two different cellular functions, one of which is cell adhesion (reviewed in Peifer, 1995), while the other is cell-to-cell signalling (reviewed in Moon *et al.*, 1997a and Orsulic and Peifer, 1996). Recently, it has been shown that the latter pathway directly affects gene expression via a transcription factor, Lef1 (Behrens *et al.*, 1996; Huber *et al.*, 1996). The potential complexity of Wnt interactions is, clearly, of great importance when considering their roles in the development of individual organs.

From the early expression studies on mouse *Wnt* genes it was clear that several of them had very specific patterns of expression in the developing central nervous system (CNS) and that they were likely to be important for normal CNS development (Gavin *et al.*, 1990; Parr *et al.*, 1993). This was supported by the phenotypes observed in naturally occurring mutants and in transgenic mice where individual *Wnt* genes had been 'knocked out'; for example, in *Wnt1*-/- transgenic mice the midbrain and cerebellum fail to develop (McMahon *et al.*, 1992) and more recent studies show that central and peripheral neuronal development is altered in mutants from as early as embryonic day 9.5 and confirm that both the midbrain and rhombomere 1 are deleted (Mastick *et al.*, 1996). Evidence from the less severe mouse mutant *swaying* (which has a point mutation in the *Wnt1* gene) indicates that *Wnt1* expression is required for the correct formation of the midbrain–hindbrain border (Thomas *et al.*, 1991). The roles of *Wnt1*, *Engrailed* and *Pax* genes in regulating the development of midbrain and hindbrain have been reviewed by Joyner (1996). Studies on transgenic mice have also shown that *Wnt* genes are important to the development of other organs, for example *Wnt4* in the developing kidney (Stark *et al.*, 1994) and *Wnt7a* in the developing limb (Parr and MacMahon, 1995).

2. Human *WNT* genes

Complete or partial clones have been isolated for 13 human *WNT* genes. In vertebrates as a whole 18 different *Wnt* genes have been identified to date, although in no single species has the full complement of genes been isolated. *Table 1* gives details of the human genes identified, their chromosomal localization, where known, and the available information about their expression patterns. As can be seen from the table, the expression of several genes has been studied in cancer tissues and cell lines, particularly from breast and colon cancers. Abnormal expression of some of the genes is associated with these malignancies and work is being carried out to understand the mechanisms involved and the role which may be played by other proteins in the Wnt pathway. As can also be seen from the table, there is very little information about the developmental expression of human *WNT* genes.

Given the already known human/mouse differences in orthologous gene expression patterns (see Chapter 2) and the relative paucity of information about gene expression in the developing human brain, a critically important organ in which we would expect there to be species differences in gene expression (see above), we decided to carry out an expression study of developmentally important genes in the embryonic human brain. We chose to examine the expression of three *Wnt* genes initially: *WNT5A* and *WNT7A*, which had been shown in mice to be expressed in the developing forebrain and to have very sharp boundaries of expression, both there and in other brain regions (Parr *et al.*, 1993), and *WNT8B* [the human gene was the first reported mammalian orthologue (Lako *et al.*, 1996)], which we have shown is also expressed in very specific regions of the forebrain as well as, weakly, in the midbrain (see below).

The expression patterns of these genes were determined using *in situ* hybridization following methods outlined by Wilson (see Chapter 7) and given in more detail in Moorman *et al.*, 1993. We studied embryos from Carnegie Stages 12 to 21, corresponding to approximately the 26–52 day period (see Chapter 3), which is equivalent to between 10/10.25 and 13.5 days post coitum in the mouse. This covers the period from just after the closing of the anterior neuropore (Carnegie Stage 11) to the establishment of the major forebrain subregions: the cerebral cortex, hippocampus, basal ganglia, thalamus and hypothalamus. The first stages of cerebral hemisphere development can be distinguished at Carnegie Stage 14 and by Carnegie Stage 21 the cortical plate is visible in some areas (O'Rahilly and Muller, 1994). *Figure 1a* is a schematic diagram of the development of the brain at Carnegie Stage 14 and shows the subdivision of the forebrain into telencephalon and diencephalon, the midbrain (mesencephalon) and the regionalization of the hindbrain to metencephalon and myelencephalon. *Figure 1b* shows a transverse

Table 1. Human *WNT* genes

Human *WNT* gene	Location	Accession number	Expression	Expression not detected in	Reference
1	12q13	X03072		Normal, breast fibroadenomas and malignant breast tissue	van Ooyen and Nusse, 1984; Huguet *et al.*, 1994
2	7q3.1	X07876	Adult lung and digestive tissue Placenta Fetal lung Breast fibroblasts Fetal lung fibroblast cell lines Overexpressed in colorectal cancer Overexpressed in breast fibroadenomas	Adult peripheral blood lymphocytes, cerebral cortex and cultured sweat duct epithelium Fetal liver, kidney, pancreas and colon	Huguet *et al.*, 1994; Vider *et al.*, 1996; Wainwright *et al.*, 1988
3	17q21		Normal, breast fibroadenomas and malignant breast tissues and some breast cancer cell lines		Huguet *et al.*, 1994; Roelink *et al.*, 1993
3A			Low levels in normal breast tissue Overexpressed in breast fibroadenomas	Normal, breast fibroadenomas and malignant breast tissue	Huguet *et al.*, 1994
4					Huguet *et al.*, 1994
5A	3p21–p14	L20861 and U39837 (promoter region)	Widespread expression in adult tissues and cell lines Embryonic expression (see text and *Table 2*) Low levels in breast cell lines and normal breast tissue Overexpressed in breast fibroadenomas and malignant breast tissue Aberrant expression in lung and prostate carcinomas and melanomas		Clark *et al.*, 1993; Danielson *et al.*, 1995; Iozzo *et al.*, 1995; LeJeune *et al.*, 1995

Clone	Locus	Accession	Expression	Negative detection	Reference
7A	3p25	U53476	Expressed in adult brain, spleen, testis, endometrial tissues and placenta; also, by RT–PCR, expression was detected in kidney, uterus, fetal lung and fetal brain. Embryonic expression (see text and *Table 3*). Expressed in endometrial carcinoma cell lines and endometrial tumours	By RT–PCR no expression was detected in adult breast, pancreas, prostate, heart, small intestine, colon, ovary or thyroid. By Northern blot expression was also absent from adult lung, liver, skeletal muscle, thymus and peripheral blood leukocytes	Bui *et al.*, 1997a; Ikegawa *et al.*, 1996
7B			Normal and benign breast tissue, overexpressed in some breast tumours. Upregulated in superficial (but not in invasive) bladder cancers		Huguet *et al.*, 1994
8B	10q24	X91940	Embryonic expression — only detected in the brain (see text and *Figure 2*)		Lako *et al.*, 1996
10B	12q13	U81787	Low levels of expression in normal and benign proliferations of breast tissue; elevated expression in some breast carcinomas		Bui *et al.*, 1997b
13	1p13	Z71621	Adult heart, brain, lung, prostate, testis, ovary, small intestine and colon. Placenta. Fetal brain, lung and kidney. MKN28 and MKN74 (gastric cancer) cell lines. HeLa (cervical cancer) cell line		Katoh *et al.*, 1996

We have also isolated partial genomic and cDNA clones for human *WNT76* (J. Rankin, unpublished observations) and *WNT11* (M. Lako, unpublished observations) and are currently mapping these and examining their expression patterns. RT–PCR, reverse transcriptase polymerase chain reaction.

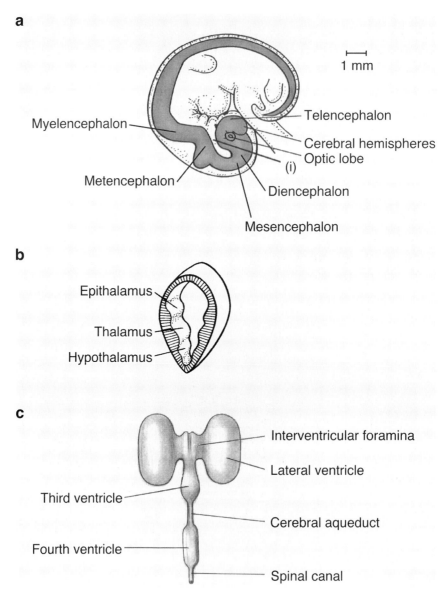

Figure 1. The major brain regions and their cavities. (a) Schematic diagram of a Carnegie Stage 14 embryo with the major brain regions indicated: forebrain–telencephalon and diencephalon; midbrain (mesencephalon) and hindbrain–metencephalon and myelencephalon. (b) Section through the line marked (i) in the diencephalon. (c) Primitive ventricles, as seen in the brain of the embryo in (a) after straightening out. The lateral ventricles are the cavities of the telencephalon, the third ventricle that of the diencephalon, the cerebral aqueduct that of the midbrain (mesencephalon) and the fourth ventricle that of the hindbrain. This figure is slightly modified from figures 12.4 and 12.5 in *Medical Embryology* by John McLachlan. Reproduced by permission of Addison Wesley Longman Ltd.

section taken at the level of the line [labelled (i) in *Figure 1a*] through the dien-cephalon. The diagram in *Figure 1c* shows the cavities (ventricles) in each of the developing brain regions. Examples of embryos between Carnegie Stages 12 and 21 and further discussion of their development at each stage are given in Chapter 3.

Our results have shown that all three *Wnt* genes studied are expressed in the germinative neuroepithelium in all areas of the brain where expression was detected and at all developmental stages examined. The germinative neuroepithe-lium (also called the ventricular layer) is the proliferative cell layer which lies nearest the ventricle in each brain region. The expression in each of the major brain regions is described in more detail below.

2.1 Expression in the developing forebrain

Of the three genes involved, *WNT8B* has the most restricted pattern of expres-sion: it is only expressed in the developing brain and within the brain it is only expressed in a limited number of regions. The more complex patterns of the *WNT5A* and *WNT7A* genes are summarized in *Tables 2* and *3*, respectively. Danielson *et al.* (1995) reported results from a smaller *in situ* hybridization study of *WNT5A* expression in human embryos between 28 and 42 embryonic days (approximately Carnegie Stages 13–17). In addition to the areas detected in our study and detailed in *Table 2*, they reported expression of *WNT5A* in part of the atrium of the heart at 28 embryonic days and transiently in the pericardium at 33 embryonic days. We have seen a weak signal only in Carnegie Stage 21 in a local-ized area of the dorsal atrium. They also reported expression in the metanephric (definitive) kidney, whereas we detected a signal in the mesonephric but not the metanephric kidney. These differences may be due to the expression being very precise in time and location in these structures, which will make any discrepan-cies in plane of section and embryonic stage of the embryos studied highly signif-icant. More detailed studies of the developing heart and kidney would be needed to resolve these points. The same study also did not find expression of *WNT5A* in the forebrain or in the rostral hindbrain as we report below. Our study of *WNT5A* expression in the brain was more extensive and, for example, we found that expression in the rostral hindbrain is confined to a very small region, the midline sulcus (see *Figure 3c*).

WNT8B is expressed in the telencephalon throughout the period under study (Carnegie Stages 15–21) in a highly regionalized manner and, at all stages, expres-sion was detected in the ventricular layer, that is in the germinative neuroepithe-lium. At Carnegie Stage 15, transcripts were detected in a domain corresponding to the region described as the hippocampal formation (O'Rahilly and Muller, 1994). From Carnegie Stage 16, the region in which expression is detected can be defined as the primordial hippocampus (*Figure 2a* and *b* illustrate the expression at Carnegie Stage 21). During Carnegie Stage 17, the primordial hippocampus is thought to be subdivided into two regions, one corresponding to the primordium gyrus dentatus and the other the primordium cornu Ammonis, the hippocampus itself (Gasser, 1975). *WNT8B* transcripts are detected throughout the whole length of the primordial hippocampus (*Figure 2a* and *b*; and M. Lako, unpublished

Table 2. *WNT5A* Expression

	Carnegie Stage 12	Carnegie Stage 15	Carnegie Stage 17	Carnegie Stage 19	Carnegie Stage 21
Forebrain	Floor of caudal diencephalon	Floor of diencephalon as continuum of midbrain expression, only as far as inframamillary recess in midline and caudal extreme of subthalamus laterally			
Midbrain	Ventral surface gives strong signal along length of midbrain — as far as sulcus limitans laterally				
Hindbrain	Rostral cord and caudal half of hindbrain ventrally. In the rostral quarter signal is confined to the midline sulcus up to the midbrain boundary				
Limb		Confined to AER and underlying mesenchyme	AER signal, and expression in mesenchymal condensations of cartilaginous anlagen		Strong at tip of digits, surrounding chondrifying phalanges and proximal radius
Gonadal ridge		Coelomic epithelium and the underlying region of in-growth, along entire length		Epithelium still strongest, but underlying mesenchyme of now discrete gonad uniformly positive, as is caudal mesonephros	
Branchial arch mesoderm / facial processes	Signal in mesodermal substance of arches		Signal becomes confined to maxillary and mandibular processes; and tongue and laryngeal areas in the midline		
Genital area			Mesenchyme underlying the genital tubercle and folds		
Herniated bowel loops		Strong signal in midgut loops in physiological umbilical hernia, from the muscular mesoderm-derived layers		Weak signal only	Not detected
Trachea			Not detected	Strong, in ventral wall only (not bronchi)	Entire length ventrally and main bronchi

Using 1100 bp cDNA clone courtesy of Professor A. Harris, Oxford.
AER, apical ectoderm ridge.

Table 3. *WNT7A* expression

	Carnegie Stage 12	Carnegie Stage 15	Carnegie Stage 17	Carnegie Stage 19	Carnegie Stage 21
Forebrain		Widespread lateral diencephalon, extending into telencephalon	Still in dorsal and ventral thalamus (but not extending into telencephalon)	Weaker and apparently less extensive in dorsal and ventral thalamus	
Midbrain		Signal in dorsolateral walls, although some extension below sulcus limitans; absent in dorsal midline area. Becomes weaker temporally			
Hindbrain		Whole length, strong signal ventrally	Whole length as far as isthmic portion of hindbrain rostrally, where signal very weak or absent. Signal overlaps sulcus limitans into dorsal domain slightly		
Spinal Cord		Ventral signal along entire length			
Limb	Ectoderm over limb bud mesenchyme dorsally	Signal in dorsal ectoderm layer	Expression not detected in differentiated limbs		

Using 1200 bp cDNA clone courtesy of Professor A. Harris, Oxford.

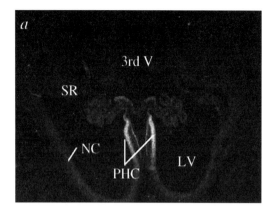

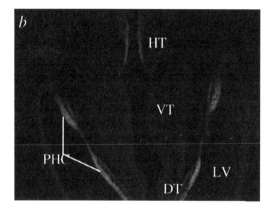

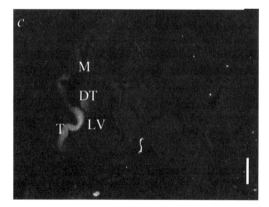

Figure 2. Expression of *WNT8B* in the human embryonic forebrain. (a) Dark-field microscopy of a transverse section (Carnegie Stage 21) hybridized to a human *WNT8B* antisense probe: a strong signal can be seen in the primordial hippocampus. Ventral is at the top of the figure and dorsal is at the bottom of the figure. (b) Dark-field microscopy of a more caudal (i.e. towards the tail) transverse section (Carnegie Stage 21) hybridized to a human *WNT8B* antisense probe: expression is detected in the hypothalamus as well as the primordial hippocampus. Ventral is at the top of the figure and dorsal is at the bottom of the figure. (c) Dark-field microscopy of a sagittal section (Carnegie Stage 15) hybridized to a human *WNT8B* antisense probe: expression is detected in part of the dorsal thalamus as well as the telencephalon [signal in primardial hippocampus, see (a) and (b)] In each case as a negative control, a human *WNT8B* sense probe was hybridized to an adjacent section and no expression was detected (data not shown). DT, dorsal thalamus; HT, hypothalamus; LV, lateral ventricle; M, midbrain; NC, neocortex; PHC, primordial hippocampus; SR, striatal ridge; T, telencephalon; VT, ventral thalamus; 3rdV, third ventricle.

observations), probably indicating a role in its differentiation from other cortical formations rather than subspecialization of the primordial hippocampus itself. During the following stages (Carnegie Stages 18–21), *WNT8B* continues to be expressed in the ventricular layer of the primordial hippocampus. The expression of *WNT*8B at Carnegie Stage 21 is shown in *Figure 2a–c*. We cannot say at the earlier stages whether the cells expressing *WNT8B* will only give rise to the primordial hippocampus. However, by Carnegie Stage 21 it is clear that while there is expression in the primordial hippocampus no expression is detected in neighbouring structures (see *Figure 2a* and M. Lako, unpublished observations). We observed an identical expression pattern in mouse embryos at 10–13 days post coitum (M. Lako, unpublished observations).

WNT5A is not expressed in the telencephalon in any of the stages studied (Carnegie Stages 12–21), while *WNT7A* is only expressed transiently in Carnegie Stage 15 rostrally to the telencephalon–diencephalon junction (*Table 3*). In the mouse, neither *Wnt5a* nor *Wnt7a* signals were detected in the telencephalon at 9.5 days post coitum (Parr *et al.*, 1993).

All three genes are expressed in the diencephalon (see *Figure 1b* for a diagrammatic representation of the major diencephalic subdivisions): *WNT5A* ventrally (in part of the hypothalamus); *WNT7A* mostly dorsolaterally (in the thalamus); and *WNT8B* in two discontinuous regions, one dorsolateral (in part of the thalamus) and one ventral (in part of the hypothalamus). During Carnegie Stage 15, the developing thalamus is seen to be separated by different sulci into epithalamus, dorsal thalamus and ventral thalamus, while the hypothalamus divides into the subthalamus and hypothalamus proper [from dorsal to ventral (O'Rahilly and Muller, 1994)]. *WNT5A* is expressed in the floor of the diencephalon, as a continuum of midbrain expression (see below), from Carnegie Stage 12 and throughout the time studied. At later stages this region can be distinguished as the subthalamic component of the hypothalamus (*Table 2* and *Figures 3a–c*). In the human, *WNT7A* is expressed dorsolaterally in the diencephalon, in the dorsal and ventral thalamus (*Table 3* and *Figure 4a* and *b*). The signal appears relatively stronger and more extensive at Carnegie Stages 15 and 17 and then becomes weaker and reduced in its extent in the dorsal and ventral thalamus at Carnegie Stages 19 and 21. In the mouse at 9.5 days post coitum, the expression of *Wnt5a* and *Wnt7a* is described as having a similar, ventral distribution, although the *Wnt5a* signal is much stronger (Parr *et al.*, 1993). This is unlike the situation in the human at later stages (Carnegie Stages 15–21, equivalent to mouse 11–13.5 days post coitum) where the expression patterns of *WNT5A* and *WNT7A* are clearly different, with the *WNT7A* expression being dorsolateral rather than ventral (compare *Figures 3a* and *4c*). We have looked at the expression of *Wnt5a* at later stages in the mouse (11–14 days post coitum) and found that its ventral expression persists and is similar to that seen in the human at comparable stages (*Figure 3a* and P. Bullen, unpublished observations). We are currently analysing the expression of *Wnt7a* in the mouse at later stages.

At Carnegie Stage 15, *WNT8B* transcripts are confined to cells in the germinative neuroepithelium of part of the dorsal thalamus (*Figure 2c*). In subsequent embryonic stages (Carnegie Stages 16–21) extension of the thalamus as well as some stratification occurs. However, the *WNT8B* expression remains restricted to the germinative neuroepithelium and is not detected over the full length of the dorsal thalamus (M. Lako, unpublished observations). In addition, we detected a continuous expression of *WNT8B* in the germinative neuroepithelium of the posterior hypothalamus dorsal (*Figures 2b* and *5b*) which will give rise to some of the future hypothalamic nuclei.

In the early stages (Carnegie Stages 15 and 16), expression of *WNT8B* is also detected in the diencephalon–telencephalon junction (M. Lako, unpublished observations).

2.2 Expression in the developing midbrain

Of the three genes under study, *WNT5A* is expressed most strongly in the midbrain. As in the diencephalon, it is expressed ventrally along the length of the

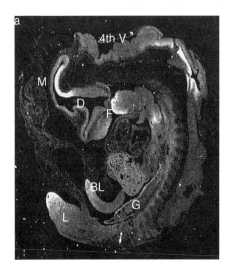
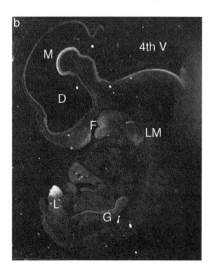
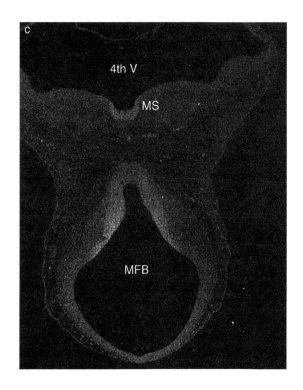

midbrain (*Figure 3a* and *3b*). In contrast, *WNT7A* is expressed dorsolaterally (*Figure 4c*), although there is some extension into the ventral region and, as in the diencephalon, there may be a small region where both genes are being expressed (*Figure 5c* and *d*). As in the diencephalon, *Wnt5a* and *Wnt7a* expression is reported to be very similar in the midbrain of the mouse at 9.5 days post coitum and is detected in a broad ventral and lateral region (Parr *et al.*, 1993). *Wnt7a* expression is reported as being weaker and patchier than that of *Wnt5a*. Again, our studies at later stages in the mouse confirm that the reported ventral distribution of *Wnt5a* persists and is similar to that found at equivalent stages in the human (*Figure 3a* and *b*). We are currently analysing the expression of *Wnt7a* in the mouse at later stages, but in the human it is clear that the expression of the two genes is very different at the stages studied (compare *Figure 3a* and *b* with *Figure 4c*).

In the midbrain, *WNT8B* is the most weakly expressed of the three genes studied. At Carnegie Stage 15 weak expression of *WNT8B* is detected along the roof of the midbrain. This expression domain could not be detected after Carnegie Stage 16. We also detected faint signals in a patch of cells in the caudal midbrain just abutting the midbrain–hindbrain border throughout the stages we studied.

2.3 Expression in the developing hindbrain

In the hindbrain, the *WNT5A* signal is more restricted than the *WNT7A* signal. The latter is detected ventrally over the whole length of the hindbrain and, indeed, ventrally over the whole length of the spinal cord (*Figure 4b* and *c*). *WNT5A*, however, is detected ventrally in the caudal half of the hindbrain and only in the rostral spinal cord, again ventrally (*Figure 3a* and *b*). In the rostral half of the hindbrain the signal is restricted to the midline sulcus (*Figure 3c*). In the hindbrain, the pattern for both genes is very similar to that seen in the mouse (Parr *et al.*, 1993 and P. Bullen, unpublished observations).

Figure 3. Expression of *WNT5A* at Carnegie Stages 15 and 17. Dark-field microscopy of sagittal sections at Carnegie Stages 15 (a) and 17 (b) hybridized to a human *WNT5A* antisense probe showing midbrain signal extending just into the caudal floor of the diencephalon, which is demonstrated as the caudal subthalamus on transverse section at Carnegie Stage 17 (c). Also shown in (a) and (b) are expression in the caudal hindbrain [floor of the 4th ventricle (4th V)], facial and laryngeal mesenchyme, coelomic epithelium of the gonadal ridge, lower limb (distal–proximal gradient) and midgut bowel loop [(a) only]. The signal in rostral (i.e. towards the head) hindbrain is seen to be limited to the ventral midline sulcus (c). In (c), because of the angle of flexion of the head, the dorsal midbrain–forebrain border (MFB) is at the bottom of the figure while the ventral midbrain–forebrain border is in the middle, just below the ventral hindbrain and the dorsal hindbrain is at the top of the figure. In each case, as a negative control, a human *WNT5A* sense probe was hybridized to an adjacent section and no expression was detected (data not shown). BL, bowel loop; D, diencephalon; F, facial mesenchyme; G, gonadal ridge; L, limb; LM, laryngeal mesenchyme; M, midbrain; MS, midline sulcus.

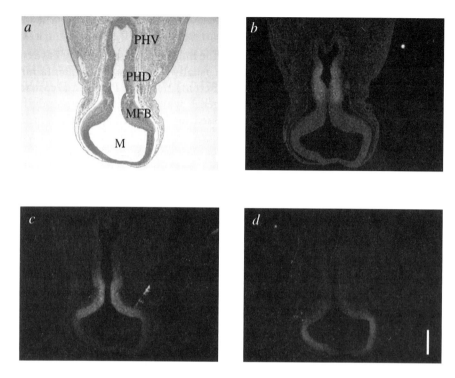

Figure 5. Apparent overlapping expression of three human *WNT* genes in the embryonic brain at Carnegie Stage 15 (approximately 33 days post-ovulation). (a) Light-field microscopy of a haematoxylin- and eosin-stained transverse section (Carnegie Stage 15). (b) Dark-field microscopy of a transverse section (Carnegie Stage 15) hybridized to a human *WNT8B* antisense probe: expression is detected in the posterior hypothalamus dorsal. (c) Dark-field microscopy of a transverse section (Carnegie Stage 15) hybridized to a human *WNT5A* antisense probe: expression is detected at the midbrain–forebrain border (d) Dark-field microscopy of a transverse section (Carnegie Stage 15) hybridized to a human *WNT7A* antisense probe: expression is detected in the midbrain. From these and other experiments (data not shown), it appears that there may be overlapping expression at the expression boundaries of these genes. Confirmation of this requires double-labelling experiments (see text). In each case appropriate sense probes were hybridized to adjacent sections as negative controls and no expression was detected (data not shown). M, midbrain; MFB, midbrain–forebrain border; PHD, posterior hypothalamus dorsal; PHV, posterior hypothalamus ventral.

layer (reviewed in Barbe, 1996). The specific role(s) which individual *WNT* genes may be playing in the regions in which they are expressed will depend on which cell types are expressing each gene and which cell types are receiving the signal within the ventricular layer. *Wnt* genes are known to be able to act in both an autocrine and paracrine manner (Jue *et al.*, 1992; Lin *et al.*, 1992) and so the identity of both expressing and receiving cell types is important. The identity of expressing cell types could be found in double-labelling experiments using anti-bodies against marker proteins indicative of the different cell types of the ventricular layer providing that the expression of individual *Wnt* genes is able to be detected non-radioactively. Of the genes that we studied, *WNT5A* and *WNT8B* should be expressed sufficiently strongly, at least in most areas of their distribution. From *Figure 5* it can be seen that *WNT5A* and *WNT7A* may be expressed in the same small region and, similarly, the expression domains of *WNT5A* and *WNT8B* may also have a small overlap. Double-labelling experiments should clarify whether a region of overlap exists and to what extent. In the mouse, we already know that more than one *Wnt* gene is expressed in some brain regions, for example *Wnt7b* and *Wnt8b* are both expressed in the primordial hippocampus (Parr *et al.*, 1993; M. Lako unpublished observations), and it will be very interesting to determine whether these genes are being expressed in the same cell type.

The identification of the target cell type(s) for individual *WNT* genes *in vivo* will depend on identification of their specific receptors. As discussed in the introduction, work is being carried out in other species to try and match individual *Wnt* genes with specific members of the *Fz* family and to see whether there are other types of receptor or other proteins involved with the Fz proteins (e.g. Axelrod *et al.*, 1996; Blair, 1996). Detailed examination of the situation in the human awaits the outcome of these experiments. Again, it will be very interesting to see whether one cell type can express more than one Wnt receptor and, if so, whether there are differences in the timing of their expression.

Two important types of information have already emerged from our studies on *Wnt* gene expression in the human which highlight the value of this kind of work, in general, as well as, more specifically, guiding our work for the future. The first is illustrated by our results on *WNT8B* expression where the patterns detected, in the stages studied, are very similar in the human to that found in the mouse. This considerably strengthens any extrapolation from mouse studies to human for this gene. Functional studies of genes during development are clearly very difficult to carry out in humans. Thus, the increased reliability that comparable expression patterns for orthologous genes give to such studies in the mouse, as a model for what is happening in the human, is very important. The functional significance of the limited expression of *WNT8B* in the diencephalon is unclear at present. Its expression pattern, however, suggests that it will be a very good molecular marker for the study of hippocampal differentiation and raises the possibility that it has a role in patterning of this region. We have mapped the *WNT8B* gene to chromosome 10q24 (Lako *et al.*, 1996) and its expression patterns suggest that the *WNT8B* gene could be the locus for certain neurological disorders in humans which arise as a result of abnormal development of the hippocampus. Indeed, one form of partial epilepsy has already been linked to markers in 10q24 (Ottoman *et al.* 1995).

The second kind of information to emerge, in the case of *WNT7A*, is that there may be differences between the expression patterns in mice and humans (see Chapter 2). Clearly, it is very important to establish the functional significance of any expression differences which may be detected. This should clarify whether the differences are meaningful and, if so, may give important clues about mechanisms underlying anatomical/functional differences between mouse and human and/or may provide information about the mode of action or range of functions of the specific gene under study. Again studies in mice will be very important for testing any proposed differences in gene function which may be hypothesized from the expression patterns.

In order to characterize the function(s) of individual genes, it is clear that knowledge about the expression patterns of interacting genes will be very important. There is an increasing need, therefore, to study in a coordinated way the expression patterns in the human embryo of networks or pathways of genes which can then be compared with results from the mouse. For these types of studies, the mouse atlas and gene expression database of mouse development (see Chapter 14) will be of crucial importance, as will establishing a similar atlas and gene-expression database for the human embryo.

Acknowledgements

We are grateful to the Wellcome Trust and the Newcastle University Hospitals Special Trustees for their support of this project. We would also like to thank Professor Adrian Harris (Institute of Molecular Medicine, Oxford) for his kind gift of the *WNT5A* and *WNT7A* cDNA clones.

References

Axelrod JD, Matsuno K, Artavanis-Tsakonas S, Perrimon N. (1996) Interactions between *wingless* and *notch* signalling pathways mediated by *dishevelled*. Science **271**: 1826–1832.

Baker NE. (1987) Molecular cloning of sequences from *wingless*, a segment polarity gene in *Drosophila*: the spatial distribution of a transcript in embryos. *EMBO J.* **6**: 1765–1773.

Barbe MF. (1996) Tempting fate and commitment in the developing forebrain. *Neuron* **16**: 1–4.

Behrens J, Kries JPV, Kuhl M, Bruhn L, Wedlich D, Grosschedl R, Birchmeier W. (1996) Functional interaction of β-catenin with the transcription factor LEF-1. *Nature* **382**: 638–642.

Bhanot P, Brink M, Samos CH, Hsieh J-C, Wang Y, Macke JP, Andrew D, Nathans J, Nusse R. (1996) A new member of the *frizzled* family from *Drosophila* functions as a *wingless* receptor. *Nature* **382**: 225–230.

Blair SS. (1996) Notch and wingless signals collide. *Science* **271**: 1822–1823.

Bui TD, Lako M, LeJeune S, Curtis ARJ, Strachan T, Lindsay S, Harris AL. (1997a) Isolation of a full-length human *WNT7A* gene implicated in limb development and cell transformation, and mapping to chromosome 3p25. *Gene* **189**: 25–29.

Bui TD, Rankin J, Smith K, Huguet EL, Ruben S, Strachan T, Harris AL, Lindsay S. (1997b) A novel human *Wnt* gene, *WNT10B*, maps to 12q13 and is expressed in human breast carcinomas. *Oncogene* **14**: 1249–1253.

Clark CC, Cohen I, Eichstetter I, Cannizzaro LA, McPherson JD, Wasmuth JJ, Iozzo RV. (1993) Molecular-cloning of the human protooncogene *WNT-5a* and mapping of the gene (*WNT5A*) to chromosome 3p14-p21. *Genomics* **18**: 249–260.

Danielson KG, Pillarisetti J, Cohen IR, Sholehvar B, Huebner K, Ng LJ, Nicholls JM, Cheah KSE, Iozzo RV. (1995) Characterization of the complete genomic structure of the human *WNT-5A* gene, functional-analysis of its promoter, chromosomal mapping, and expression in early human embryogenesis. *J. Biol. Chem.* **270**: 31225–31234.

Dickinson ME, Krumlauf R, Mcmahon AP. (1994) Evidence for a mitogenic effect of *Wnt-1* in the developing mammalian central-nervous-system. *Development* **120**: 1453–1471.

Gasser RF. (1975) *Atlas of Human Embryos*. Harper & Row, Hagerston, MA.

Gavin BJ, McMahon JA, McMAhon AP. (1990) Expression of multiple *Wnt-1/int-1-* related genes during fetal and adult mouse development. *Genes Dev.* **4**: 2319–2332.

Hollyday M, Mcmahon JA, McMahon AP. (1995) *Wnt* expression patterns in chick-embryo nervous-system *Mech. Dev.* **52**: 9–25.

Huber O, Korn R, McLaughlin J, Ohsugi M, Hermann BG, Kemler R. (1996) Nuclear localisation of β-catenin by interaction with transcription factor LEF-1. *Mech. Dev.* **59:** 3–10.

Huguet EL, McMahon JA, McMahon AP, Bicknell R, Harris AL. (1994) Differential expression of human *WNT* gene-2, gene-3, gene-4, and gene-7b in human breast cell-lines and normal and disease states of human breast-tissue. *Cancer Res.* **54:** 2615–2621.

Ikegawa S, Kumano Y, Okui K, Fujiwara T, Takahashi E, Nakamura Y. (1996) Isolation, characterization and chromosomal assignment of the human *WNT7A* gene. *Cytogenet. Cell Genet.* **74:** 149–152.

Iozzo RV, Eichstetter I, Danielson KG. (1995) Aberrant expression of the growth-factor *WNT-5A* in human malignancy. *Cancer Res.* **55:** 3495–3499.

Joyner AL. (1996) *Engrailed, Wnt* and *Pax* genes regulate midbrain–hindbrain development. *Trends Genet.* **12:** 15–20.

Jue SF, Bradley RS, Rudnicki JA, Varmus HE, Brown AMC. (1992) The mouse *Wnt-1* gene can act via a paracrine mechanism in transformation of mammary epithelial cells. *Mol. Cell Biol.* **12:** 321–328.

Katoh M, Hirai M, Sugimura T, Terada M. (1996) Cloning, expression and chromosomal localization of *WNT-13*, a novel member of the wnt gene family. *Oncogene* **13:** 873–876.

Kelly GM, Greenstein P, Erezyilmaz DF, Moon RT. (1995) Zebrafish *wnt8* and *wnt8b* share a common activity but are involved in distinct developmental pathways. *Development* **121:** 1787–1799.

Lako M, Strachan T, Curtis ARJ, Lindsay S. (1996) Identification and characterisation of *WNT8B*, a novel human *Wnt* gene which maps to 10q24. *Genomics* **35:** 386–388.

Lejeune S, Huguet EL, Hamby A, Poulsom R, Harris AL. (1995) *WNT5-alpha* cloning, expression, and up-regulation in human primary breast cancers. *Clin. Cancer Res.* **1:** 215–222.

Lin TP, Guzman RC, Osborn RC, Thordarson G, Nandi S. (1992) Role of endocrine, autocrine and paracrine interactions in the development of mammary hyperplasia in *Wnt-1* transgenic mice. *Cancer Res.* **52:** 4413–4419.

Marin-Padilla M. (1988) Early ontogenesis of the cerebral cortex. In: *Cerebral Cortex Vol. 7: Development and Maturation of the Cerebral ortex* (eds A Peters, EG Jones). Plenum Press, New York, NY, pp. 1–34.

Mastick GS, Fan CM, Tessier Lavigne M, Serbedzija GN, McMahon AP, Easter SS Jr. (1996) Early deletion of neuromeres in *Wnt*1(-/-) mutant mice: evaluation by morphological and molecular markers. *J. Comp. Neurol.* **374:** 246–258.

Mayr T, Deutsch U, Kuhl M, Drexler CA, Lottspeich F, Deutzmann R, Wedlich D, Risau W. (1997) Fritz: a secreted frizzled-related protein that inhibits Wnt activity. *Mech. Dev.* **63:** 109–125.

McMahon AP, Joyner AL, Bradley A, McMahon JA. (1992) The midbrain hindbrain phenotype of *Wnt1-/-* mice results from stepwise deletion of *engrailed*-expressing cells by 9.5 days postcoitum. *Cell* **69:** 581–595.

Moon RT, Brown JD, Torres M. (1997a) *WNTs* modulate cell fate and behaviour during vertebrate development. *Trends Genet.* **13:** 157–162.

Moon RT, Brown JD, Yang-Snyder JA, Miller JR. (1997b) Structurally related receptors and antagonists compete for secreted Wnt ligands. *Cell* **88:** 725–728.

Moorman AFM, De Boer PAJ, Vermeulen JLM, Lamers WH. (1993) Practical aspects of radio-isotopic in situ hybridisation on RNA. *Histochem. J.* **25:** 251–266.

O'Rahilly R, Muller F. (1994) *The Embryonic Human Brain: an Atlas of Developmental Stages.* Wiley-Liss, New York, NY.

O'Rourke NA, Chenn A, McConnell SK. (1997) Postmitotic neurons migrate tangentially in the cortical ventricular zone. *Development* **124:** 997–1005.

Orsulic S, Peifer M. (1996) Cell-cell signaling: Wingless lands at last. *Curr. Biol.* **6:** 1363–1367.

Ottoman R, Risch N, Hauser WA et al. (1995) Localisation of a gene for partial epilepsy to chromosome 10q. *Nature Genetics* **10:** 56–60.

Parr BA, McMahon AP. (1995) Dorsalizing signal *Wnt-7a* required for normal polarity of D-V and A-P axes of mouse limb. *Nature* **374:** 350–353.

Parr BA, Shea MJ, Vassileva G, McMahon AP. (1993) Mouse *Wnt* genes exhibit discrete domains of expression in early embryonic CNS and limb buds. *Development* **119:** 247–261.

Peifer M. (1995) Cell adhesion and signal transduction: the Armadillo connection. *Trends Cell Biol.* **5:** 224–229.

Rakic P. (1988) Specification of cerebral cortical areas. *Science* **241:** 170–176.

Rakic P. (1995) A small step for the cell, a giant leap for mankind: a hypothesis of neocortical expansion during evolution. *Trends Neurosci.* **18**: 383–388.

Rijsewijk F, Schuermann M, Wagenaar E, Parren P, Weigel D, Nusse R. (1987) The *Drosophila* homologue of mouse mammary oncogene int-1 is identical to the segment polarity gene *wingless*. *Cell* **50**: 649–657.

Roelink H, Wang J, Black Dm, Solomon E, Nusse R. (1993) Molecular-cloning and chromosomal localization to 17q21 of the human *WNT3* gene. *Genomics* **17**: 790–792.

Stark K, Vainio S, Vassileva G, McMahon AP. (1994) Epithelial transformation of metanephric mesenchyme in the developing kidney regulated by *Wnt-4*. *Nature* **372**: 679–683.

Thomas K, Musci T, Neumann P, Capecchi M. (1991) *Swaying* is a mutant allele of the protooncogene *Wnt-1*. *Cell* **67**: 969–976.

Van Ooyen A, Nusse R. (1984) Structure and nucleotide sequence of the putative mammary oncogene *int-1*; proviral insertions leave the protein-encoding domain intact. *Cell* **39**: 233–240.

Vider BZ, Zimber A, Chastre E *et al.* (1996) Evidence for the involvement of the *WNT-2* gene in human colorectal-cancer. *Oncogene* **12**: 153–158.

Wainwright BJ, Scambler PJ, Stanier P *et al.* (1988) Isolation of a human gene with protein sequence similarity to human and murine *int-1* and the *Drosophila* segment polarity mutant *wingless*. *EMBO J.* **7**: 1743–1748.

12

Fragile X syndrome: new insights into Fragile X mental retardation syndrome from research on human developing tissues

Marc Abitbol

1. Introduction

The Fragile X mental retardation syndrome is characterized by moderate mental retardation, a long face with large, everted ears and macro-orchidism. It is an X-linked disorder and the most frequent heritable cause of mental retardation in humans (Reiss *et al.*, 1994 and references therein). Fragile X syndrome usually results from the expansion of the CGG repeat in the *FMR1* gene. There is strong evidence that *FMR1* gene alterations play a central role in the clinical syndrome. Amplification of the trinucleotide repeat is associated with hypermethylation of the CpG island 58 to *FMR1* (Bell *et al.*, 1991; Heitz *et al.*, 1991; Oberlé *et al.*, 1991; Vincent *et al.*, 1991) and a marked decrease or silencing of *FMR1* transcription (Pieretti *et al.*, 1991). Additional evidence of a unique role for *FMR1* in the pathogenesis of the Fragile X mental retardation syndrome comes from two patients with a partial or complete deletion of the *FMR1* gene who exhibit Fragile X syndrome in the absence of the fragile site (Gedeon *et al.*, 1992; Wohrle *et al.*, 1992) and the discovery of a single point mutation in the open-reading frame of the *FMR1* gene in a patient with very severe Fragile X syndrome, but without cytogenetic expression of the fragile site (De Boulle *et al.*, 1993). Together this evidence strongly suggests that the *FMR1* gene is directly responsible for Fragile X syndrome.

Molecular Genetics of Early Human Development, edited by T. Strachan, S. Lindsay and D.I. Wilson.
© 1997 BIOS Scientific Publishers Ltd, Oxford.

Developmental studies in both mice (Hinds *et al.*, 1993) and humans (Abitbol *et al.*, 1993) revealed *FMR1* gene expression patterns compatible with the phenotypic variability observed in affected patients as well as with the major manifestations of this disease. The precocious and strong *FMR1* expression in developing and adult neurons in the human (Abitbol *et al.* 1993; Devys *et al.*, 1993; C. Agulhou, A. Kobetz, A. Sittler, J.-L. Mandel, A. Malafosse and M. Abitbol, unpublished information) is consistent with the occurrence of mental retardation. The intense *FMR1* transcription in developing chondrocytes of normal human embryos and fetuses can be reconciled with the characteristic facial phenotype of Fra(X) boys and with the limb abnormalities reported in some Fra(X) patients. The strong *FMR1* expression in testis during germ cell proliferation may relate to the development of macro-orchidism in affected boys and adult males (Bächner *et al.*, 1993a,b; Devys *et al.*, 1993; Malter *et al.*, 1997).

Fmr1 knockout mice (the Dutch–Belgian Fragile X Consortium, 1994) lack normal *Fmr1* RNA and protein and show enlarged testes, impaired cognitive function and aberrant behaviour. Comery and his colleagues have recently shown that dendritic spines on apical dendrites of layer V pyramidal cells in the occipital cortex of *Fmr1* knockout mice are longer than those in wild-type mice and that they are often thin and tortuous (Comery *et al.*, 1997). This parallels the neuropathological findings in the brains of Fra(X) patients and indicates that FMR1 protein (FMRP) expression may be required for normal spine morphological development. Furthermore, spine density along the apical dendrite is greater in the knockout mice and the authors suggested that this may reflect impaired developmental organizational processes of synapse stabilization and elimination or pruning (Comery *et al.*, 1997).

The combination of these data is sufficient to validate mutations in the *FMR1* gene as the sole cause of Fragile X mental retardation syndrome, but the data do not provide insights concerning the molecular mechanisms underlying the diverse manifestations of this puzzling disorder, nor any information about the mechanisms of the triplet repeat expansion which is the origin of the disease in most instances.

FMR1 shows a tissue-specific variation in its level of expression that correlates well with the morphological and neuropsychological consequences of Fragile X syndrome (Abitbol *et al.*, 1993; Devys *et al.*, 1993; Hinds *et al.*, 1993) but the major result emerging from the developmental studies of *FMR1* gene expression, including both Northern blot and immunohistochemical analyses of embryonic, fetal and adult tissues, is the ubiquitous expression of *FMR1* transcripts and proteins (FMRPs).

It is difficult to reconcile the absence of obvious, detectable, histological alterations in most adult tissues from patients with Fragile X syndrome with this ubiquitous gene expression pattern and protein distribution. One clue to the resolution of this apparent discrepancy comes from the observation that the *FMR1* premessenger RNA can undergo alternative splicing. As a result of alternative splicing of the 17 exons present in the *FMR1* gene, several isoforms shorter than the full length, 614 amino acids, form of FMRP are detected *in vivo* (Ashley *et al.*, 1993b; Sittler *et al.*, 1996; Verkerk *et al*, 1993). The significance of these isoforms remains unclear and some of them may be specifically and/or predominantly expressed in distinct tissues or within particular cellular subpopulations. *In situ* hybridization on tissue sections, including human embryonic tissue sections, using specific oligonucleotide probes corresponding to each particular exon may help to clarify this issue.

To date, the *FMR1* gene expression pattern has been studied only in human fetuses of 8 and 9 weeks. Here, we report preliminary results obtained in our laboratory on *FMR1* gene expression in human embryos of 3 to 6 weeks, as well as on human fetal tissues of 8, 9 and 25 weeks. The use of distinct oligonucleotide probes hybridizing specifically to transcripts containing exon 14 [presence or absence of which has been shown to correlate with nuclear or cytoplasmic localization of the FMRP (Sittler *et al.*, 1996)] or to *FMR1* transcripts without exon 14, allowed us to establish the tissue distribution of these *FMR1* transcripts from very early stages of human development.

2. Materials and methods

2.1 Tissue preparation

Morphologically normal human embryos ($n = 14$), ranging from 3 to 6 postovulatory weeks were obtained from legal medical abortions using Mifepristone (RU 486) at the Hôpital Broussais in Paris. Normal human fetuses were obtained from legal abortions carried out for medical reasons concerning the health of the pregnant women. Complete independence was maintained between the medical staffs of the hospital and the research group. Maternal consent to the research was never asked for before the abortion was decided on and performed. The consent was obtained after the abortion was completed. This procedure was approved by the Ethical Committee of the Hôpital Necker-Enfants Malades (Paris). The embryos were microdissected from the whole trophoblast under a dissecting microscope. The developmental stage of each embryo was assessed following the Carnegie classification established by O'Rahilly and Muller (1987; see also Chapter 3, for a description of the major developmental features of embryos at each of the stages studied). The experiments were carried out on 14 embryos and four fetuses: two Carnegie Stage 10 embryos (approximately 22 days of development); one Carnegie Stage 11 embryo (approximately 24 days of development); two Carnegie Stage 13 embryos (approximately 28–32 days of development); two Carnegie Stage 15 embryos (approximately 33 days of development); two Carnegie Stage 16 embryos (approximately 37 days of development); two Carnegie Stage 17 embryos (approximately 41 days of development); two Carnegie Stage 18 embryos (approximately 44 days of development); one Carnegie Stage 20 embryo (approximately 51 days of development); one 8-week and one 9-week human fetus, as well as on two 25-week human fetal brains. Fetal tissues and microdissected embryos were frozen using powdered ice, and stored at –80°C. Cryostat sections (15 μm) were mounted on slides previously coated with 2% 3-aminopropyl-triethoxylane solution in acetone. Sections were fixed for 30 min in 2% paraformaldehyde in 0.1 M phosphate buffer (pH = 7.4), rinsed once in phosphate buffered saline, rinsed briefly in water and dehydrated through a series of ethanols of increasing concentrations (50%, 75%, 95%). Sections were then air dried and finally stored at –80°C. This procedure was used to preserve the mRNAs of embryonic and fetal tissues.

2.2 DNA probes for in situ hybridization

The oligonucleotide probes were synthesized and purified by Genset, France. They were 3′ end labelled with a ^{35}S dATP (NEN) using terminal deoxyribonucleotidyl transferase (BRL, 15 U ml^{-1}) at a specific activity of approximately 7×10^8 cpm µg^{-1}. The probes were purified on biospin columns (Biorad) before use. The Genbank accession numbers for the *FMR1* sequences are: L29074, L38501, M67468 and X61378.

The sequence of the *FMR1* common antisense probe (FMRcas) is:

5′-GACCATCCAC GCTGTCTGGC TTCTCTTTCT TCTGGTTACG ATCTTTACCC GTGCGCAGCC -3′.

The sequence of the common *FMR1* sense probe (FMRcs) is:

5′-GGCTGCGCAC GGGTAAAGAT CGTAACCAGA AGAAAGAGAA GCCAGACAGC GTGGATGGTC-3′.

The sequence of the exon 14 *FMR1* antisense probe (*FMR14*as) is:

5′- ATATCCAGGA CCGCGTCTGC CGTGCCCCCT ATTTCTGTAA GGTCTACTAC -3′.

The sequence of the exon 14 *FMR1* sense probe (FMR14s) is:

5′- GTAGTAGACC TTACAGAAAT AGGGGGCACG GCAGACGCGG TCCTGGATAT -3′.

The sequence of the exons 13–15 *FMR1* antisense probe (FMR13/15as) is:

5′-CGCCGCCGTC CGTCTCCTCT GCGCAGGAAG CTCTCCCTCT CTTC-CTCTGT TGGAGCTTTA AATAGTTCAG GTGATAATCC -3′.

The sequence of the exons 13–15 *FMR1* sense probe (FMR13/15s) is:

5′- GGATTATCAC CTGAACTATT TAAAGCTCCA ACAGAGGAAG AGAGGGAGAG CTTCCTGCGC AGAGGAGACG GACGGCGGCG -3′.

The sense probes were used as negative controls and a rat arrestin oligonucleotide was also used as a negative control. The Genbank accession number for the rat arrestin cDNA is M60737 and the sequence of the oligonucleotide used is:

5′- GTCTCTCTTC CCCAGGTAGA TGGTCACCGA CTTGTCCCGG GAGACCTTCT TGAAGATGAC -3′.

All the sequences of the oligonucleotide probes were chosen by means of the Oligo software in order to discard autocomplementary hybridizations, loops and hairpins corresponding to potential stable structures. Additionally, the oligonucleotide sequences were tested against all the available sequences of the GenBank database using the Fasta and Blast procedures.

2.3 In situ hybridization procedure

The hybridization cocktail contained: 50% formamide, 4 × SSC (standard saline citrate), 1 × Denhardt's solution; 0.25 mg ml^{-1} yeast tRNA; 0.25 mg ml^{-1} sheared herring sperm DNA; 0.25 mg ml^{-1} poly-A (Boehringer Mannheim); 10% dextran sulphate (Sigma); 100 mM dithiothreitol; and a ^{35}S dATP-labelled probe, at a concentration of 6×10^5 cpm 100 μl^{-1} of final hybridization volume. One hundred microlitres of hybridization solution was put on each section, then the sections were covered with a parafilm coverslip and incubated in a humidified chamber at 43°C for 20 h. After hybridization, the slides were washed twice in 1 × SSC containing 10 mM DTT for 15 min each at 55°C, and twice in 0.5 × SSC containing 10 mM DTT for 15 min each at 55°C, and finally in 0.5 × SSC containing 10 mM DTT for 15 min at room temperature. The sections were then dipped in water, dehydrated through a series of graded concentrations of ethanol as before and exposed to Amersham Hyperfilm betamax X-ray films for 4 days and then to Kodak NTB2 photographic emulsion for 2 months at +4°C.

2.4 Northern blot hybridization

The tissue-specific pattern of *FMR1* expression was assessed by hybridization of the three oligonucleotide antisense probes (*FMR*cas, *FMR*14as and *FMR*13/15as), labelled with T4 polynucleotide kinase (Life Technologies), to Northern blot filters containing human poly A$^+$ RNAs (2 μg for each lane) from different fetal and adult tissues (Clontech, Palo Alto: references 7750-1, 7755-1, 7759-1, 7760-1, 7756-1). mRNAs from the following adult tissues were present on the Clontech blots: amygdala; caudate nucleus; corpus callosum; hippocampus; substantia nigra; subthalamic nucleus; thalamus; cerebellum; cerebral cortex; medulla; occipital pole; frontal lobe; temporal lobe; putamen; spinal cord; spleen; thymus; prostate; testis; ovary; small intestine; colon (lining); peripheral blood leukocytes; heart; whole brain; placenta; lung; liver; skeletal muscle; kidney; and pancreas. mRNAs from the following fetal tissues were present on the Clontech blots: brain; liver; lung; and kidney. Hybridization and washing were carried out as follows: each blot was incubated overnight (20 h) with 5 ml of prehybridization buffer containing 50% formamide, 5 × SSC, 50 mM Na$_2$HPO$_4$ (pH 6.5), dextran solution × 1 (Sigma), 250 mg ml^{-1} salmon sperm DNA and 10 μg ml^{-1} poly(A) (Boehringer-Mannheim). Hybridization was performed for 20 h at 42°C in the same solution except that 50 mg ml^{-1} salmon sperm DNA and 10^5 cpm per cm^2 of labelled probe were added. Each membrane was washed twice with 100 ml of 2 × SSC/0.1% SDS (sodium dodecyl sulphate) each for 30 min, once with 0.3 × SSC/0.1% SDS and once with 0.1 × SSC/0.1% SDS for 30 min at 65°C. The membrane was then subjected to autoradiography with X-OMAT X-ray film (Kodak) with an intensifying screen at –70°C.

3. Results

Northern blot hybridizations to filters containing mRNAs from both adult and fetal tissues (Clontech; see Section 2), with the three different antisense oligonucleotides as probes (*FMR*cas, *FMR14*as, *FMR*13/15as) detected transcripts of the same size: a major band of 4.4 kb is consistently detected in all tissues tested (data not shown). Thus the *FMR*cas oligonucleotide probe, which should detect all *FMR1* transcripts, detected the same autoradiographic signals as those obtained when using oligonu-

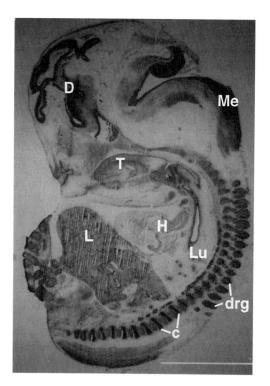

Figure 1. Cresyl violet coloured parasagittal section of a 9-week-old human fetus. This section is adjacent to the one on which *FMR1* signals were visualized by autoradiography (see *Figure 2*). c, cartilage of vertebral bodies; D, diencephalon, drg, dorsal root ganglion; H, heart; L, liver; Lu, lung; Me, medulla; T, tongue. (Scale bar = 5 mm.)

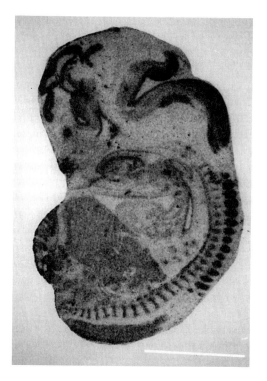

Figure 2. Autoradiogram of a 9-week-old fetal tissue section hybridized with the *FMR*cas probe (Scale bar = 5 mm.)

cleotide probes designed to distinguish between transcripts containing exon 14 (*FMR*14as) and those without exon 14 (*FMR*13/15as) (data not shown). Similarly, whichever probe was used, we detected an ubiquitous but non-uniform *FMR1* gene expression pattern in developing embryos. From the earliest stage examined (Carnegie Stage 10), the *FMR1* gene is expressed at highest abundance in the proliferating neurons of the central nervous system and in the hepatocytes of the developing liver. The *FMR1* gene is also already expressed in mesenchymal cells which contribute to developing cartilage, although at a lower intensity (C. Agulhou, A. Kobetz, A. Sittler, J.-L. Mandel, A. Malafosse and M. Abitbol, unpublished information). *FMR1* gene expression is lower still in the developing lungs and blood vessels. It is, however, clearly detectable in the developing myocardial cells (data not shown).

In 8- and 9- week-old fetuses, autoradiographic and cellular hybridization signals are prominent in the central nervous system (especially in the neural retina) and in mesenchyme cells (particularly in cartilage). Less intense signals are observed in the liver, lung and in other organs (*Figures 1, 2* and *3*).

In 25-week-old fetal brain tissue sections, the three antisense oligonucleotide probes (*FMR*cas, *FMR*14as, *FMR*13/15as) detect *FMR1* mRNAs expressed at highest abundance in the large pyramidal cells of the hippocampus (see *Figure 6*) and in the cholinergic neurons of the nucleus basalis magnocellularis. The *FMR1* gene is expressed in all proliferating, migrating and differentiating neurons observed at this fetal stage. Cerebral cortex, caudate nucleus and thalamus (*Figures 4* and *5*) as well as subthalamic nuclei and substantia nigra appear strongly

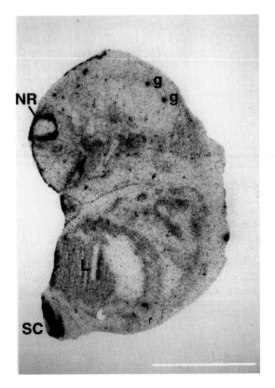

Figure 3. Autoradiogram of a 9-week-old fetal lateral parasagittal tissue section hybridized with the *FMR*cas probe. The retinal signal is clearly visible. g, ganglion; L, liver; NR, neural retina; SC, caudal part of spinal cord (Scale bar = 5 mm.)

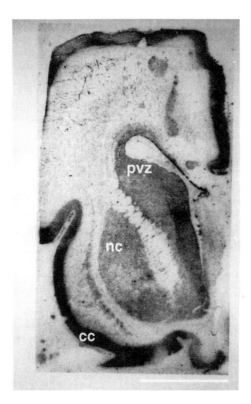

Figure 4. Autoradiogram of a 25-week-old fetal frontal cerebral tissue section hybridized with the *FMR*cas probe. cc,cerebral cortex; nc, caudate nucleus; pvz, periventricular zone. (Scale bar = 1 cm.)

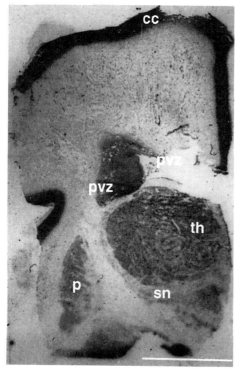

Figure 5. Autoradiogram of a 25-week-old fetal frontal cerebral tissue section hybridized with the *FMR*cas probe. cc, cerebral cortex; p, putamen; pvz, periventricular zone; th, thalamus; sn, subthalamic nucleus. (Scale bar = 1 cm.)

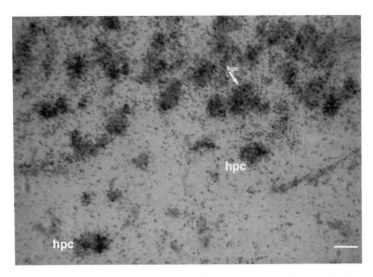

Figure 6. Very intense *FMR1 in situ* hybridization cellular labelling of the human fetal pyramidal cells of hippocampus at 25 weeks. hpc, hippocampal pyramidal cells. (Scale bar = 100 μm.)

labelled at the autoradiographic levels. We cannot exclude the possibility that both glial cells and neurons express the *FMR1* gene and FMRP. However, the corpus callosum does not appear to be significantly labelled, which would suggest that there is no expression of the *FMR1* gene in glial cells. Double-labelling experiments would be required to address this question directly.

4. Discussion

Fragile X syndrome is caused by a deficiency of FMRP, but little is known about the physiological role of this novel gene product. *FMR1* mRNAs and FMRP are expressed in many early human developing and adult human tissues (*Figures 2–6* and data not shown) and strong expression is detected in several regions of the brain, as has been detailed in the present study.

FMRP is found predominantly in the cell cytoplasm, although occasional nuclear staining has been reported (Devys *et al.*, 1993; Verheij *et al.*, 1993). Two human genes closely related to *FMR1*, termed *FXR1* and *FXR2* have recently been identified (Siomi *et al.*, 1995; Zhang *et al.*, 1995). FMRP has been reported to form specific homomultimers as well as heteromultimers with FXR1 and FXR2 proteins, when expressed in yeast (Zhang *et al.*, 1995). Whether this happens in human cells remains to be demonstrated, although it is tempting to speculate that the stoichiometric content of such complexes may be critical for cell physiology, particularly in neurons and primordial germ cells of testis, both of which are known to have an extremely high level of transcriptional complexity. Our preliminary study provides qualitative results which demonstrate that there is no obvious tissue segregation or tissue-specific expression of two distinct *FMR1* mRNA populations (that is +/– exon 14). It does not rule out the possibility that differences may exist at a quantitative level. Such quantitative differences in the levels of *FMR1* splicing isoforms expression may be involved in the process of multimerization.

One clue to the function of FMRP originates from the identification of sequences homologous to two motifs existing in several RNA-binding proteins (Ashley *et al.*, 1993b; Musco *et al.*, 1996; Siomi *et al.*, 1993a,b). FMRP contains two sequence elements that show homology to the heterogeneous nuclear ribonucleoprotein K homology (KH) domain and one sequence element that is comparable to the so-called RGG box. The KH module exists in single or multiple copies in a diverse set of proteins, the only common property of which is that all function in close association with RNA. KH-containing proteins play major roles in the regulation of cellular RNA metabolism. FMRP can indeed bind RNA with some degree of sequence specificity *in vitro* (Ashley *et al.*, 1993a; Siomi *et al.*, 1993b), although no *in vivo* specific RNA target has yet been identified. The identification of a missense mutation in the second FMRP KH domain, which blocks *in vitro* RNA binding (De Boulle *et al.*, 1993; Siomi *et al.*, 1994) and causes a severe form of Fragile X syndrome, is clearly consistent with the hypothesis that focuses on RNA binding as the major function for FMRP. The predominant cytoplasmic localization of FMRP in expressing cells combined with these data leads to the suggestion that FMRP may play a role in regulating the translation of specific target mRNA species *in vivo* (Ashley *et al.*, 1993a; Khandjian *et al.*, 1996; Siomi *et al.*, 1993b).

Recently, several groups have demonstrated that FMRP is associated with cytoplasmic ribosomes, which is clearly consistent with a role for FMRP in the regulation of translation. Tamanini *et al.* (1996) demonstrated also that FMRP is associated with ribosomes via RNA. Weiler *et al.* (1997) reported that *FMR1* mRNA rapidly associates with synaptic polyribosomal complexes in synaptoneurosomes (synaptic subcellular fractions) after stimulation by a specific metabotropic glutamate receptor agonist. Moreover, immunostaining of the synaptosomal proteins at short intervals after stimulation shows increased FMRP expression relative to unstimulated samples, indicating rapid synthesis of FMRP in response to synaptic activation. Earlier, it had been shown more generally that protein translation occurs in developing synapses in response to neurotransmitter stimulation. These data, combined with those obtained by Comery *et al.* (1997), strongly suggest that fast production of FMRP near synapses in response to activation may be important for the normal maturation of synaptic connections.

A novel minor isoform of FMRP has been reported which lacks exon 14 sequences (Sittler *et al.*, 1996). Exon 14 contains a protein nuclear export signal which is functionally identical to the nuclear export signal of the Rev regulatory protein encoded by human immunodeficiency virus (HIV) type 1 (Fischer *et al.*, 1995; Fridell *et al.*, 1996; Meyer *et al.*, 1996; Wen *et al.*, 1995). In HIV 1 Rev, this nuclear export signal fulfils a critical role in the Rev-mediated nuclear export of RNA molecules which are bound by the Rev RNA binding domain (Malim *et al.*, 1989, 1991). Bardoni *et al.* (1997) further dissected the FMRP nuclear export signal and nuclear localization signal domains: three leucine residues in exon 14 are functionally important for the cytoplasmic localization of FMRP and the nuclear localization signal activity is localized between residues 115–150 in a region that lacks stretches of basic residues which are typical of most nuclear localization signals. Willemmsen *et al.* (1996) also demonstrated an association of FMRP with ribosomal precursor particles in the nucleolus.

Together these data suggest an unexpected nuclear role for the normally cytoplasmic FMRP which may function to permit the sequence-specific nuclear export of target RNA molecules in expressing cells (Fridell *et al.*, 1996). FMRP may have a role similar to that of ribosomal RNA-binding proteins which are synthesized in the cytoplasm, enter the nucleus, to be assembled into ribosomes in the nucleolus and are then returned to the cytoplasm as a ribosomal RNP (ribonucleoprotein). A clear prediction of this model is that translation, if not transcription, of multiple mRNA species should be perturbed in target tissues for which *FMR1* expression appears to be critical for normal function (e.g. brain and testis). Extensive experiments performed in our laboratory on normal and mutated Fragile X fetal brain tissues, currently, do not confirm this prediction (data not shown).

There are still many puzzling questions to be answered about the role(s) of *FMR1* RNA and FMRP; for example, the major question whether *FMR1* is involved only in translation or in both transcription and translation remains to be resolved. In doing so questions whether *FMR1* RNA is associated with the complex splicing machinery in cells, whether FMRP is a structural component of ribosomes and whether *FMR1* RNA is a structural component of spliceosomes or ribosomes will need to be addressed. The powerful tools of structural biology are now allowing us to visualize, in detail, ribosomes and spliceosomes, with all their components, and this should enable us to answer these questions and to overcome the difficulties of interpretation which arise from unexpected cross-hybridization of oligonucleotide probes and misleading cross-reactions of antibodies.

5. Conclusion

Initially, developmental studies of *FMR1* gene expression allowed the scientific community to validate further *FMR1* as the gene responsible for Fragile X mental retardation and to eliminate the hypothesis that this syndrome is a contiguous gene syndrome. Further studies, reported here, have revealed that *FMR1* plays a major role in human development, from the earliest stages examined (Carnegie Stage 10, approximately 22 embryonic days). A comparative three-dimensional reconstruction of *FMR1* gene expression patterns, as detected by *in situ* hybridization on human embryonic serial tissue sections, using a panel of specific oligonucleotide probes may help to elucidate the still enigmatic function(s) of both *FMR1* mRNAs and FMRP.

The need for further *FMR1* expression mapping experiments on human developing tissues has been highlighted by a recent study of Malter *et al.* (1997) on testes of 13-week and 17-week gestation fetuses carrying full *FMR1* mutations and ovaries from a female fetus of 16 weeks' gestation carrying the full *FMR1* mutation on one of her two X chromosomes. This elegant study disproves the hypothesis that the germline is protected from full expansion of the triplet repeat and suggests that there can be full mutation contraction in the immature testis. Thus, full expansion may already exist in the maternal oocyte, or if postzygotic, it may arise quite early in development, prior to germline segregation from somatic lineages. Clearly, immunohistochemistry and *in situ* hybridization on appropriate 3-week- to 8-week-old embryonic tissue sections may help to clarify this issue.

References

Abitbol M, Menini C, Delezoide A-L, Rhyner T, Vekemans M, Mallet J. (1993) Nucleus basalis magnocellularis and hippocampus are the major sites of FMR-1 expression in the human fetal brain. *Nature Genetics* **4**: 147–152.

Ashley CT Jr, Wilkinson KD, Reines D, Warren ST. (1993a) FMR1 protein: Conserved RNP family domains and selective RNA binding. *Science* **262**: 563–566.

Ashley CT, Sutcliffe JS, Junst CB, Leiner HA, Eichier EE, Nelson DL, Warren ST. (1993b) Human and murine FMR-1: alternative splicing and translational initiation downstream of the CGG repeat. *Nature Genetics* **4**: 244–251.

Bächner D, Steinbach P, Wohrle D, Just W, Vogel W, Hameister H, Manca A, Poustka A. (1993a) Enhanced FMR-1 expression in testis. *Nature Genetics* **4**: 115–116.

Bächner D, Manca A, Steinbach P, Wohrle D, Just W, Vogel W, Hameister H, Poustka A. (1993b) Enhanced expression of the murine *FMR1* gene during germ-cell proliferation suggests a special function in both the male and the female gonad. *Hum. Mol. Genet.* **2**: 2043–2050

Bardoni B, Sittler A, Shen Y, Mandel JL. (1997) Analysis of domains affecting intracellular localisation of the FMRP protein. *Neurobiol. Diseases* (in press).

Bell MV, Hirst MC, Nakahori Y et al. (1991). Physical mapping across the fragile-X – hypermethylation and clinical expression of the Fragile-X syndrome. *Cell* **64**: 861–866.

Comery TA, Harris JB, Willems PJ, Oostra BA, Irwin SA, Weiler IJ, Greenough WT. (1997) Abnormal dendritic spines in fragile X knockout mice: maturation and pruning deficits. *Proc. Natl Acad. Sci. USA* **94**: 5401–5404.

De Boulle K, Verkerk AJMH, Reyniers et al. (1993) A point mutation in the FMR-1 gene associated with fragile X mental retardation. *Nature Genetics* **3**: 31–35.

Devys D, Lutz Y, Rouyer N, Bellocq J-P, Mandel J-L. (1993) The FMR-1 protein is cytoplasmic, most abundant in neurons and appears normal in carriers of a fragile X premutation. *Nature Genetics* **4**: 335–340.

Fischer U, Huber J, Boelens WC, Mattaj lW, Lührmann SR. (1995) The HIV-I Rev activation domain is a nuclear export signal that accesses an export pathway used by specific cellular RNAs. *Cell* **82**: 475–483.

Fridell RA, Fischer U, Luhrmann R, Meyer BE, Meinkoth JL, Malim MH, Cullen BR. (1996) Amphibian transcription factor IIIA proteins contain a sequence element functionally equivalent to the nuclear export signal of human immunodeficiency virus type I Rev. *Proc. Natl Acad. Sci. USA* **93**: 2936–2940.

Gedeon AK, Baker E, Robinson H et al. (1992). Fragile X syndrome without CCG amplification has an *FMR1* deletion. *Nature Genetics* **1**: 341–344.

Heitz D, Rousseau F, Devys D et al. (1991) Isolation of sequences that span the Fragile-X and identification of a Fragile-X related CpG island. *Science* **251**: 1236–1239.

Hinds HL, Ashiey CT, Sutoliffe JS, Nelson DL, Warren ST, Housman DE, Schalling M. (1993) Tissue specific expression of FMR-1 provides evidence for a functional role in fragile X syndrome. *Nature Genetics* **3**: 36–43.

Khandjian EW, Corbin F, Woerly S, Rousseau F. (1996) The fragile X mental retardation protein is associated with ribosomes. *Nature Genetics* **12**: 91–93.

Malim MH, Hauser J, Le SY, Maizel JV, Cullen BR. (1989) The HIV-1 Rev transactivator acts through a structured target sequence to activate nuclear export of unspliced viral mRNA. *Nature* **338**: 254–257.

Malim MH, McCarn DF, Tiley LS, Cullen BR. (1991) Mutational definition of the human immunodeficiency virus type I Rev activation domain. *J. Virol.* **65**: 4248–4254.

Malter HE, Iber JC, Willemsen R, De Graaf E, Tarleton JC, Leisti J, Warren ST, Oostra BA. (1997) Characterization of the full fragile X syndrome mutation in fetal gametes. *Nature Genetics* **15**: 165–169.

Meyer BE, Meinkoth JL, Malim MH. (1996) Nuclear transport of human immunodeficiency virus type 1, visna virus, and equine infectious anemia virus Rev proteins: Identification of a family of transferable nuclear export signals. *J. Virol.* **70**: 2350–2359.

Musco C, Stier C, loseph C, Morelli MAC, Nilges M, Gibson TJ, Pastore A. (1996) Three-dimensional structure and stability of the KH domain: Molecular insights into the fragile X svndrome. *Cell* **85**: 237–245.

Oberlé I, Rousseau E, Heitz D, Kretz C, Devys D, Hanauer A, Boue J, Bertheas MF, Mandel JL. (1991) Instabilily of a 550 base pair DNA segment and abnormal methylation in fragile X syndrome. *Science* **252**: 1097–1102.

O'Rahilly R, Muller F. (1987) *Developmental Stages in Human Embryos Including a Revision of Streeter's 'Horizons' and a Survey of the Carnegie Collection.* Carnegie Institution of Washington, Washington D.C., Publication No. 637.

Pieretti M, Zhang FP, Fu Y-H, Warren ST, Oostra BA, Caskey CT, Nelson DL. (1991) Absence of expression of the *FMR1* gene in fragile-X syndrome. *Cell* **66**: 817–822.

Reiss AL, Kazazian HH, Krebs CM, Mcaughan A, Boehm CD, Abrams MT, Nelson DL. (1994) Frequency and stability of the Fragile-X premutation. *Hum. Mol. Genet.* **3**: 393–398.

Siomi H, Matunis MJ, Michael WM, Dreyfuss G. (1993a) The pre-mRNA binding K protein contains a novel evolutionarily conserved motif. *Nucleic Acids Res.* **2t**, 1193–1198.

Siomi H, Siomi MC, Nussbaum RL, Dreyfuss G. (1993b) The protein product of the fragile X gene, *FMR1*, has characteristics of an RNA-binding protein. *Cell* **74**: 291–298.

Siomi H, Choi M, Siomi MC, Nussbaum RL, Dreyfuss G. (1994) Essential role for KH domains in RNA binding: impaired RNA binding by a mutation in the KH domain of FMR I that causes the fragile X syndrome. *Cell* **77**: 33–39.

Siomi MC, Siomi H, Sauer WH, Srinivasan S, Nussbaum RL, Dreyfuss G. (1995) FXR1 an autosomal homologue of the Fragile X mental retardation gene. *EMBO J.* **14**: 2401–2408.

Sittler A, Devys D, Weber S, Mandel J-L. (1996) Alternative splicing of exon 14 determines nuclear or cytoplasmic localisation of fmrl protein isoforms. *Hum. Mol. Genet.* **5**:.95–102.

Tamanini F, Meijer N, Verheij C, Willems PJ, Galjaard H, Oostra BA, Hoogeven AT. (1996) FMRP is associated to the ribosomes via RNA. *Hum. Mol. Genet.* **5**: 809–813.

The Dutch–Belgian Fragile X Consortium (1994) Fmrl Knock out mice: a model to study Fragile X Mental retardation. *Cell* **78**: 23–33.

Verheij C, Bakker CE, de Graaff E *et al.* (1993) Characterization and localization of the FMR-I gene product associated with fragile X syndrome. *Nature* **363**: 722–724.

Verkerk AJMH, de Graaff E, De Boulle K *et al.* (1993) Alternative splicing in the fragile X gene *FMR1*. *Hum. Mol. Genet.* **2**: 399–404.

Vincent A, Heitz D, Petit C, Kretz C, Oberle I, Mandel JL. (1991) Abnormal pattern detected in Fragile-X patients by pulsed-field gel-electrophoresis. *Nature* **349**: 624–626.

Weiler IJ, Irwin SA, Klintova AY *et al.* (1997) Fragile X mental retardation is translated near synapses in response to neurotransmitter activation. *Proc. Natl Acad. Sci. USA* **94**: 5395–5400.

Wen W, Meinkoth JL, Tsien RY, Taylor SS. (1995) Identification of a signal for rapid export of proteins from the nucleus. *Cell.* **82**: 463–473.

Willemsen R, Bontekoe C, Tamanini F, Galjaard H, Hoogeven A, Ben Oostra. (1996) Association of FMRP with ribosomal particles in the nucleolus. *BBRC* **225**: 27–33.

Wohrle D, Kotzot D, Hirst MC *et al.* (1992) A microdeletion of less than 250 kb, including the proximal part of the FMR-1 gene and the fragile X-site, in a male with the clinical phenotype of fragile-X syndrome. *Am. J. Hum. Genet.* **S1**: 299–306.

Zhang Y, O'Connor JP, Siomi MC, Srinivasan S, Dutra A, Nussbaum RL, Dreyfuss G. (1995) The fragile X mental retardation syndrome protein interacts with novel homologs FXRI and FXR2. *EMBO J.* **14**: 5358–5366.

(i)   Data entry can be time consuming and requires a combination of computing and embryological expertise which is difficult to find in one individual.
(ii) Specially written custom software may only be effective in the hands of its authors and the logistics of user support may be complex.
(iii) Sophisticated computer hardware can be prohibitively expensive and the data files may be difficult to export to other hardware platforms.
(iv) Requirements for fast processing and extensive memory can mean that interactive viewing is only possible on advanced workstations.
(v)   The magnitude of the task is such that only one reference embryo at each stage may be produced.

The strategy we chose to adopt had two major aims. First we intended that models should be produced on personal computers, using inexpensive commercial software, so that they could be viewed by researchers without access to specialized hardware. Commercial software for 3-D modelling and animation developed for use in the areas of graphic design and computer-generated virtual reality significantly extends the versatility of earlier personal computer-based approaches to 3-D modelling (e.g. Keri and Ahnelt, 1991). Secondly, production of 3-D models should be relatively rapid so that *any* embryo or *particular* embryonic structure can be reconstructed. This approach would permit the visualization of unique abnormal embryos as well as reference embryos and so would lend itself to the display of particular gene expression patterns, as they become established (Wilkinson and Green, 1990).

This paper describes the methodology we employed to achieve these aims and is illustrated by reference to a study of heart development based on serially sectioned human embryos. Our technique can readily be adapted by other scientists who lack an advanced computing background to produce useful interactive 3-D models.

2. Methods

The histological sections of embryos were obtained from either the Walmsley Collection of Human Embryos, School of Biological and Medical Sciences, University of St Andrews or from the Boyd Collection of Human Embryos at the Department of Anatomy, University of Cambridge. The Walmsley and Boyd collections contain valuable historical specimens at rare embryonic stages [from 1.8 mm crown–rump length (Carnegie Stage 10) upwards]. The majority of sectioned material had been fixed, embedded in wax and stained without special regard to the provision of registration marks. We have combined the records of these two collections to form the British Universities Human Embryo Database which is accessible via the Internet (http: //embryos.st-andrews.ac.uk/). World Wide Web browsers can be used to implement interactive access to this database (Aiton *et al.*, 1997).

2.1 QuickTime VR

The term virtual reality (VR) describes a range of experiences which enable a person to interact with and explore a spatial environment through a computer. Often,

VR applications require specialized hardware or accessories, such as powerful graphics workstations, stereo displays, 3-D goggles or gloves. Apple Computer QuickTime VR™ system software allows users to experience these kinds of spatial interactions on relatively inexpensive desktop computers. It is an integral part of Apple's QuickTime system architecture and does not require additional hardware. QTVR has two components: a panoramic movie technology that allows users to explore 3-D spaces, and an object movie technology that allows users to examine objects interactively. The QTVR Authoring Tools package is proprietary software that allows the creation of interactive animations of QTVR Objects (e.g. computer-generated 3-D models), which can be manipulated in 3-D space. Production of QTVR Objects from the histological sections of the embryonic hearts was divided into five main stages:

(i)   Paper tracing and registration
(ii)   Digital scanning
(iii)   Production of anatomically and topologically correct vector graphics of individual sections
(iv)   3-D modelling
(v)   Rendering and QTVR Object production.

It was crucial that an embryologist was available at each stage of the process in order to ensure that the appropriate histological structures were traced, modelled and visualized.

Paper tracing and registration. A vertical projection microscope was used to project images of the histological sections onto a baseboard. The outline of the embryonic heart was traced in pencil. Associated structures and additional fiducial features (such as the notochord, the ventral floor plate of the neural tube, the oesophagus, the lung buds and the pericardial cavity) were also added to the tracings to aid subsequent registration (*Figure 1*). Since tissue section thickness was normally 7–10 μm, it was neither feasible, nor essential, to trace every section in order to construct the final model. In practice, the section interval was varied to ensure that rapid changes in local morphology were adequately resolved.

Pencil tracings of the embryonic heart were photocopied onto transparent A4 sheets and registration was carried out by visual inspection, using the key landmarks duplicated through the individual sections as positional markers. The notochord was selected as a fixed registration point in each section since, in practice, this gave the maximum concordance with all the other registration features. The transparencies were rotated about the notochord to maximize the match with other features. The aligned stack of transparencies was then pierced through to provide a second registration point, which was transferred directly onto the original tracings. Section identification numbers, scale bars and inter-section spacing bars were also added to the tracings in order to aid identification and help later positioning of the section data in 3-D space.

Digital scanning. The original paper tracings, with both registration marks, were digitized by scanning on a Umax PowerLook Pro II flat bed scanner at a resolution of 150 dpi with Umax Magicscan software (version 2.2.1a). Each scan was

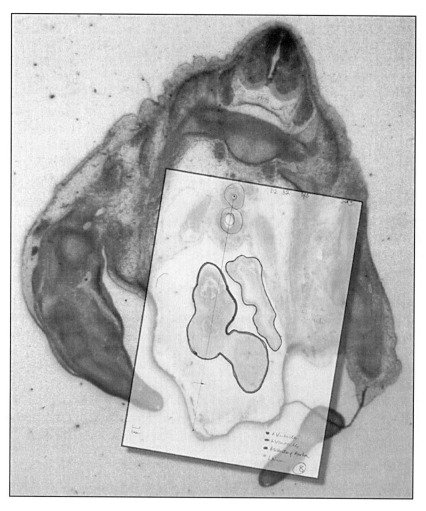

Figure 1. An embryo section projected onto a baseboard for tracing and colour coding. The larger structure outlined represents the outflow region of the heart, while the smaller is the left atrial appendage. Registration marks and lines are also visible.

saved as a 2-bit indexed colour PICT file from Adobe Photoshop. The PICT files were imported to the background layer of the drawing package Macromedia FreeHand (version 5.5), where they served as templates for outlining the structures of interest.

Production of anatomically and topologically correct vector graphics of individual sections. Structures were subsequently converted into vector graphics by using the pen tool of Macromedia FreeHand to define their outline. It was necessary to plan which regions of the vector-based FreeHand drawings would be used to represent distinct areas of the embryonic structures. Not only was it important to define where there were separate anatomical structures in each section, but also, less obviously, it was essential to consider where *topological* changes occurred in the finished model. This was particularly important when modelling complex structures which joined or branched. Initially, the anatomical and topological relationships were designated by drawing coloured lines over printed copies of the digitized line drawings. Different colours were used to distinguish the boundaries of clearly separate anatomical and topological regions. These colours were then assigned to the polygonal line segments in the FreeHand drawings to colour-code different regions of the model. This was essential for the subsequent 3-D linking of the appropriate regions of the models. When defining vertices on object outlines, care had to be taken both to maintain the same number of vertices and also to associate specific vertices with key points from one section to the next. These considerations were found to lead to improved modelling at later stages of the process. After this redrawing of the traced sections, the resulting coloured polygonal line segment drawings were exported from FreeHand as Adobe Illustrator 3.0 files, this being a suitable file format for import into the 3-D modelling package.

3-D Modelling. The completed Illustrator 3.0 files were imported to the 3-D modelling package, Strata StudioPro Blitz (version 1.75+). Scale bars, drawn on the original tracings to represent the spacing between sections, were used to create an 'armature' on which a set of planes was arranged to reflect accurately the separation of the original histological sections. The polygonal lines and registration marks for each section were then re-grouped and the armature used to locate each section in the correct position (*Figure 2*). After aligning the registration marks of the 2-D drawings in wire frame view, similarly colour-coded lines in contiguous sections were linked and skinned to produce a 3-D wire frame model (*Figure 3*). Skinning is a semi-automated procedure which constructs a polygon net connecting a line segment in one section to the equivalent line segment in the next section. The process is analogous to applying a fabric skin over the wooden ribs of an aircraft wing.

Rendering and QuickTime VR Object production. After the model was completely skinned, the rendering engine within Strata StudioPro was used to visualize the model from different views and in different colours in order to preview the positions and forms of the relevant structures. The rendering process also involves placing 'cameras' and light sources around the model to produce high quality images. A variety of rendering settings can be employed, producing images of increasing quality at the expense of extending the computing time required.

A QTVR Object is an interactive animation which can be manipulated in space. In order to make a QTVR Object in Strata StudioPro, views were rendered according to a predefined sequence of camera positions which produced a series of individual images or frames. Typically, it was found that adequate coverage was accomplished using a vertical pan of 180°, composed of seven camera arcs. On each arc, the camera was moved around 360° of horizontal pan in 24 steps (*Figure 4*). Each series of frames was saved as a QuickTime Movie prior to the construction of the QTVR Object. Frame sequences were edited together, using

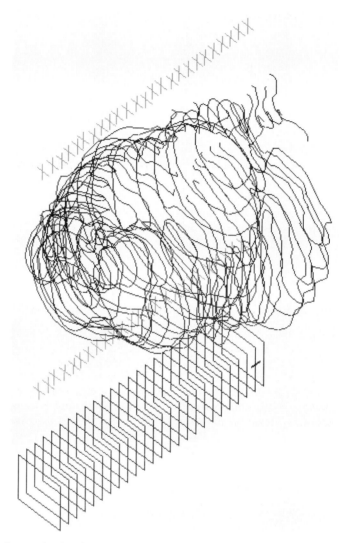

Figure 2. Isometric view from Strata StudioPro Blitz showing the armature (squares) used to space the sections, the registration marks (crosses) and the heart section outlines (polygonal lines).

Adobe Premiere 4.0. and the navigable data, which converts a QuickTime Movie into a QTVR Object, was added using Apple's Navigable Movie Player software. A QTVR Object created in this way can be viewed subsequently using Apple's QTVR Player, which allows the model to be manipulated interactively in three dimensions. A customized control interface was produced by adapting 'NavMovie XCMD Stack', a HyperCard stack included in the QTVR Authoring Tools Suite 1.0 development software (see *Figure 5*). QTVR Object files of the embryonic hearts were typically 10 to 30 Mb in size, although this was, of course, dependent upon the final pixel dimensions of the object movie. QTVR Objects can be viewed interactively on equipment of lower specification than that

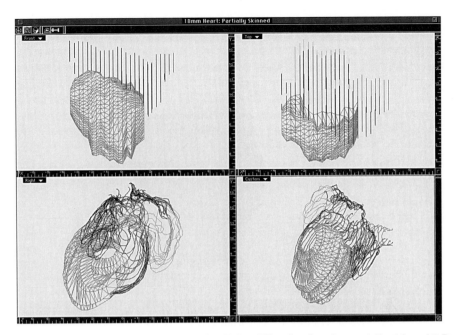

Figure 3. Modelling windows in Strata StudioPro Blitz showing the partially skinned 3-D wire frame model of an 18 mm embryo heart (Carnegie Stage 19); contiguous sections of the ventricles have been skinned using a polygon net connecting a line segment in one section to a line segment in the next section. The window labels 'front', 'top', 'right' and 'custom' relate to the orientation of the model. Since this embryo was sectioned transversely from head to tail, 'front' is cranial, 'top' is dorsal and 'right' is left with regard to the axis of the embryo. The 'custom' view can be selected by the operator to show a desired view.

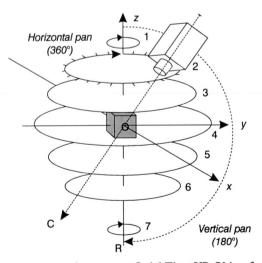

Figure 4. Camera movements used to create a QuickTime VR Object from a 3-D model. 1–7 represent the camera paths; C is an example of the camera axis; R is the axis of rotation; and O is the QuickTime VR Object (3-D reconstructed heart model) placed in the centre of the view.

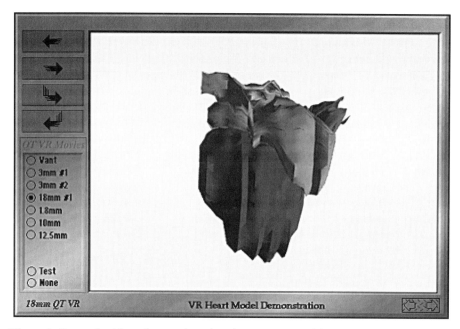

Figure 5. Customized interface used to view the reconstructed heart models produced by adapting the 'NavMovie XCMD' stack, included as part of the QuickTime Authoring Tools Suite 1.0 development software. The animations are stored as external resources to the stack and can be displayed by selecting from the extensible list. The 3-D model can be rotated around any axis by simply dragging the mouse in the appropriate direction, giving complete interactive control to the viewer.

required to generate the files (minimum requirements; Mac LCII or Windows 3.1/386 PC-compatible). Since QTVR viewers are freely available for Macintosh, Windows (3.1 and '95) and Netscape Navigator 2.0, it is simple to import VR Object files to other computer platforms.

2.2 Computing hardware and software

2-D Image acquisition, 3-D modelling and QTVR Object production were all carried out on a PowerMac 9500 with a 2 gigabyte hard disk, 96 megabytes of RAM, and a Radius ThunderColor 30/1152 accelerated QuickDraw PCI video card. A Umax PowerLook Pro II flat-bed scanner was used to scan the original tracings. Software used included Adobe Photoshop (image acquisition and editing), Adobe Premiere (movie editing), Macromedia FreeHand (vector-based graphics drawing) and Strata StudioPro (3-D reconstruction and animation). All these software packages are available from a wide range of commercial suppliers at educational discount prices. The QTVR Authoring Tools Suite was purchased through The Apple Developers Association, Apple Computer USA Ltd, Cupertino, California. The QTVR Player for both Macintosh and Windows platforms, used to replay QTVR Movies, can be downloaded from the Apple's QTVR Internet server (http://qtvr.quicktime.apple.com/).

3. Results and discussion

The aim of this work was to develop a rapid and relatively simple method of producing interactive 3-D models of early human embryonic structures from histological sections. We have successfully produced a number of these models which meet our objectives and validate our general strategy (e.g. see *Figure 5*). These are available for inspection and downloading from our World Wide Web site; the URL is http: //www.st-and.ac.uk/~www_bms/terrapin/Freebies.html

A number of issues arose during the production of the models; for example we found that the maximum number of sections which could be conveniently handled at any one time was approximately 50. Above this number, the drawing complexity and the performance of the current version of the software was such that rendering became unacceptably slow. However, it is possible to divide larger models into smaller units for modelling purposes and these more manageable units can be reassembled subsequently .

A further issue related to the processing time required to render the models. Although StudioPro has its own pre-scripted QTVR camera which shoots a sequence of frames around the object, this script resulted in excessively large movie files and extended rendering times (> 7 days). For this reason, we developed our own VR Object production protocol with a reduced number of camera positions. Nevertheless, the production of QTVR Objects is still a lengthy process (from 1 to 6 days) because of the number of frames which must be generated. As is the case with other 3-D modelling packages, the complexity of the model and choice of rendering method (solid model, Phong, Gourad, ray tracing, etc.) has a significant effect on rendering times. Anticipated improvement in processor speed may soon make these considerations less important.

One difficulty, which requires further discussion, is the surface irregularity of the models. Structures which we believed to be intrinsically smooth showed a degree of unevenness (*Figure 5*). Initially, it was assumed that this was due to inaccuracies in the alignment of the original tracings because much of the available sectioned embryo material was prepared without registration marks and we had to perform visual alignment to register the sections. Although it is possible that such an approach could lead to the jagged appearance of the model, we believe that this is unlikely. When we traced histological sections that had registration grooves on the edge of the tissue block, we still observed considerable positional shifts from section to section.

We did find that the inter-section interval was an important factor in determining the outline shape of the model, particularly at regions of significant structural change. However, even when every section was taken and the registration concordance was high, some unevenness in the model outline was still apparent. We suggest a more likely explanation of the unevenness is that the process of sectioning introduces a degree of deformation which varies from section to section depending upon the nature of the tissues. In a related project, using the same techniques and embryo collection, but involving the reconstruction of 3-D models of rigid cartilaginous structures, markedly less unevenness was observed. This tends to suggest that the amount of deformation may be related to the physical nature of the tissue sectioned. Such an interpretation is consistent with our observation that the greatest deformation is seen in sections of soft tissues surrounding hollow cavities such the early embryo heart.

The 3-D modelling software offers a choice of 'spline skin surfacing' or 'polygon skin surfacing' algorithms. Spline skin surfacing reduced the unevenness of models but sometimes generated unpredictable and misleading artefacts (*Figure 6*). Joining structures in adjacent sections with polygon skin surfacing (single straight lines) is less visually pleasing, but avoids the introduction of such artefacts. The amount of unevenness was small in relation to the absolute size of the sections. The maximum apparent displacement of one section relative to another was approximately 20 μm, and was generally much less, in a section of size 10 x 4 mm.

4. Conclusion

Whilst there are a number of very powerful computer-based 3-D reconstruction methods available, many of these still remain outside the budget and expertise of traditional embryologists. The low-cost personal computing hardware and commercial graphics software we have chosen to employ have proved highly satisfactory for our requirements although there are some restrictions placed on modelling capabilities in terms of the speed, size and structural complexity. However, the use of QTVR software offers some significant advantages when reconstructing embryonic hearts from histological sections. With QTVR, we are able to produce virtual reality objects which can be manipulated on a variety of desktop computers. The existence of appropriate cross-platform viewers means

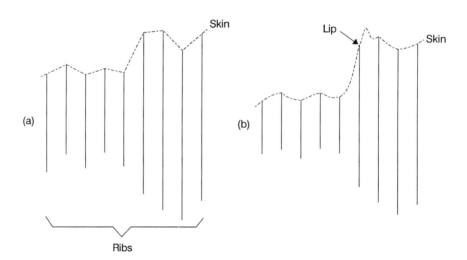

Figure 6. (a) The result of applying a polygonal skin to the outer surface of the ribs of a model. The surface that is produced is angular but there are no unusual drawing artefacts. (b) The result of applying a spline skin to the outer surface. The smoother surface skin which is produced using this method may appear more natural, although unexpected and potentially confusing artefacts can occur as demonstrated by the appearance of a 'lip' on the model surface.

that QTVR files can be viewed on Macintosh and Windows machines without the need for complex file translation. This greatly enhances both the research and teaching potential of these 3-D models.

The ability to manipulate the model in three dimensions is a valuable research tool. Although the software does not produce a true VR experience, but is rather more akin to the 'deferred virtual reality' described by Lucas et al. (1996), the speed at which the model responds to viewer control is, nonetheless, sufficient to provide an interactive interface. The recent release, QTVR 2.0, has additional features such as 'zoom' and 'translation' capabilities to allow the viewer to examine particular areas in greater detail. VR objects will have hot spots which allow the viewer to effect a transition either to a VR panorama or to another object or to another part of the same object. We anticipate that when we incorporate these improvements, it will greatly enhance the VR interface when using QTVR for 3-D reconstructions. Whilst the methods that we have described are of general applicability to visualization of morphology, they also lend themselves readily to the display of gene expression data within solid objects.

Acknowledgements

This work was supported by the British Heart Foundation Project Grant PG/94175. We gratefully acknowledge the support of the Anatomical Society, which funded the curation of the Walmsley Collection and the setting up of the British Universities Human Embryo Database, and of the Scottish Higher Education Funding Council, which funded the purchase of the Web Server through the Metropolitan Area Networks Initiative. We also gratefully acknowledge the help of Professor David Brynmor Thomas (University of St Andrews) in facilitating access to the Walmsley Collection, and Dr Visvan Navaratnam (University of Cambridge), who was responsible for creating the original Boyd Collection Database and who kindly allowed us to incorporate this material into the joint Database.

References

Aiton JF, McDonough A, McLachlan JC, Smart SD, Whiten SC. (1997) World Wide Web Access to the British Universities Human Embryo Database. *J. Anat.* 190: 149–154.

Doyle MD, Ang C, Raju R, Klein G, Williams BS, De Fanti T, Goshtasby A, Grzescuk R, Noe A. (1993) Processing cross-sectional image data for reconstruction of human developmental anatomy from museum specimens. *SIGBIO Newsletter* (Special Interest Group for Biomedical Computing of the Association for Computer Machinery) 13/1.

Keri C, Ahnelt PK. (1991) A low cost computer aided design (CAD) system for 3D-reconstructions from serial sections. *J. Neurosci. Meth.* 37: 241–250.

Lucas L, Gilbert N, Ploton D, Bonnet N. (1996) Visualisation of volume data in confocal microscopy: comparison and improvements of volume rendering methods. *J. Microsc.* 181: 238–252.

Ringwald M, Baldock R, Bard J, Kaufman M, Eppig JT, Richardson JE, Nadeau JH, Davidson D. (1994) A database for mouse development. *Science* 265: 2033–2034.

Verbeek FJ, deGroot MM, Huijsmans WH, Lamers WH, Young IT. (1993) 3D Base: A geometrical database system for the analysis and visualisation of 3D-shapes obtained from parallel serial sections including three different geometrical representations. *Comput. Med. Imaging Graph.* **17**: 151–163.

Wilkinson DG, Green J. (1990) *In situ* hybridisation and the three-dimensional reconstruction of serial sections. In: *Postimplantation Mammalian Embryos: A Practical Approach* (eds AJ Copp, DL Cockroft). Oxford University Press, Oxford, pp. 155–171.

A 3-D atlas and gene-expression database of mouse development: implications for a database of human development

Duncan Davidson and Richard Baldock

1. Introduction

We are at the beginning of an exciting period which promises a new and deeper understanding of the development of the human embryo. Recent progress has been driven by advances in molecular genetics and developmental biology which have begun to bring together two levels of description, the genotype and phenotype, at first in model organisms, particularly the mouse, and now in the human embryo itself.

Following the sequencing of the human genome, the key activity in this field will be the investigation of gene function. Already, using the mouse as a model system, a large effort is invested in producing information on gene expression during embryo development, on the phenotypic effects of mutations, and on the molecular interactions between the products of different genes. The relatively few such studies that have been made in the human embryo herald a much larger amount of work over the next decade. Several difficulties remain, however, which will limit our understanding, despite this wealth of information.

The first of these difficulties is the lack of a detailed description of the phenotype of the mammalian embryo, an essential component of genetic experiments. Unlike those of many organisms, mammalian embryos are inaccessible and, even if removed from the mother, are too opaque to allow the internal tissues to be visualized without histological sectioning. One promising approach is nuclear magnetic

Molecular Genetics of Early Human Development, edited by T. Strachan, S. Lindsay and D.I. Wilson.
© 1997 BIOS Scientific Publishers Ltd, Oxford.

resonance (NMR) microscopy (Jacobs and Frazer, 1994; *e-ref. 1*), which allows internal tissues to be visualized in unfixed, unsectioned material and may, in the future, allow visualization of probes, even in living embryos. At present, however, this method achieves a resolution far short of that obtained by histological techniques, and the potential for assaying numbers of gene-expression patterns in the same embryo is likely to be much more limited than in sectioned material, where different gene-expression patterns can be visualized in successive sections. This is an important consideration where material is scarce, as in human studies. There have, of course, been many detailed histological studies of mammalian development, but these have been carried out *ad hoc*, and have not produced a unified description which can be used as a reference framework in which to investigate gene function. Such a description is required, for example, to compare the expression of different genes or assay primary defects in the development of mutant embryos. Currently, each investigation of mutant phenotype must include a comparative study of the corresponding aspects of normal development, but the number of developmentally interesting mouse mutants is already large and will be greatly increased by mutagenesis screens in the future (Bedell *et al.*, 1997a,b; Brown and Peters, 1996).

Most histological studies have not addressed the need to visualize the three-dimensional (3-D) aspects of morphogenesis. This is especially pertinent to the study of early embryos where structures in the process of forming have ill-defined boundaries, and where proximity of different structures may regulate their differentiation. Computer-based techniques of 3-D reconstruction are now quite commonly employed, but most of these use outlines of structures to produce a wire-frame model or a surface representation derived from it. Thus, although they display the 3-D relationships between components, these reconstructions give no indication of histological structure. More importantly, they display only what the originator has chosen to outline and do not, therefore, provide a description of the embryo phenotype that is of general utility.

The alternative approach is to build 3-D reference models of representative normal embryos based on images of histological structure (*Figure 1*). All the internal tissues of such a model embryo can readily be visualized. By digitally resectioning the model in the appropriate plane, the researcher can view normal histological structure corresponding to a section through an experimental specimen. This enables the standard model to be used, for example, for comparison with many different mutant embryos. Moreover, since every volume element in the picture is, in effect, a 'pigeon hole' in a database, such a model can hold the very large amount of data — on cell activities, tissue structure, cell lineage, and so on — that is required to describe the embryo phenotype. In the following section we describe work to build an atlas of mouse development (the 'Mouse Atlas') comprising digital model embryos at successive stages of development.

A second difficulty encountered in studying mammalian development is the very large amount of information documenting gene-expression patterns in the mouse embryo. The quantity of this data, which is both temporally and spatially complex, presents serious difficulties for conventional publishing and for the individual researcher who wants to analyse his results in relation to published studies. A major use of gene-expression data is to identify gene products with overlapping or complementary distributions in the embryo in order to build a picture of potential mol-

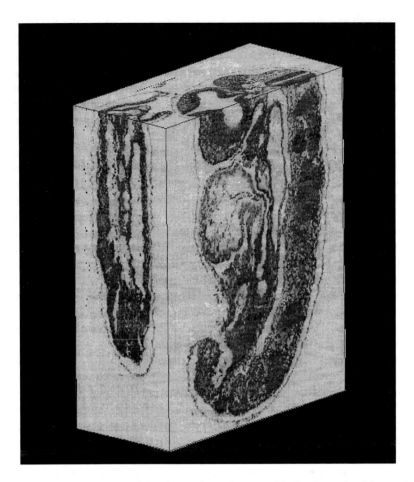

Figure 1. A 3-D reference model embryo, shown here as a block of grey-level image data. The block, reconstructed from digital pictures of successive histological sections, is composed of a 3-D array of volume elements (voxels). The model records the histological structure of the embryo, and can be resectioned in any plane as shown here. The model can also be used as the spatial framework for a database by matching data to particular groups of voxels.

ecular interactions. Moreover, since different developmental systems often make use of common pathways, even experts with a detailed knowledge of one system need to scan data from other systems for relevant information, but the sheer volume of published data is making this ever more difficult. The solution is clearly to build a database of gene-expression patterns and this is now being undertaken for several model organisms (reviewed by Davidson *et al.*, 1997a). Later in this chapter we describe work to build a gene-expression database for the mouse which is linked to the Mouse Atlas (Baldock *et al.*, 1992; Ringwald *et al.*, 1994).

Gene-expression patterns are, however, only part of the information required to understand gene function. A third problem will be to link this information to evidence, from genetic and biochemical studies, on the interactions between gene prod-

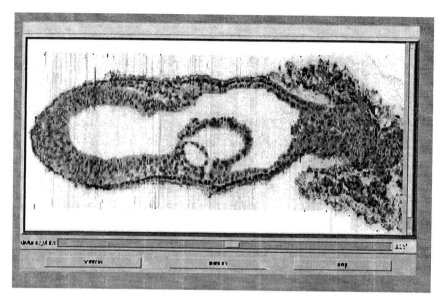

Figure 2. The Theiler Stage 10 model embryo of the Mouse Atlas. The model, originally reconstructed from transverse plastic sections, is shown here in parasagittal view. In this embryo model, the voxels represent $2 \times 2 \times 2$ μm and the model shows tissue structure in cellular detail.

custom-built software that employs an interface like the familiar 'screen-painting' packages, and data can then be assigned to this domain. The entire picture space is a three-dimensional array of voxels which can be segmented into any number of arbitrarily defined domains. Thus, many different data can be independently mapped into the same space; the model defines the visible, biological context for this information.

Data can be attached to the reference model using different modes of mapping (*Figure 4*); for example, experimental data (images or textual information on cell properties, lineage, etc.) can be attached as a block to the appropriate domain in the model. The data might be a simple bibliographic reference or a series of images of mutant embryos attached to a domain that indicates the parts of the normal embryo affected by the mutation, the 'domain of influence' of the mutation. A more sophisticated spatial mapping of images, including 3-D reconstructions, can be achieved by an image transformation process that maps each voxel (or small set of voxels) from an image of the experimental embryo to the equivalent voxels in the reference model. (Such a transformation process is an established part of the reconstruction of model embryos where it is used to compensate for local distortions that arise from sectioning.) Data from different sources can be mapped in the same reference frame thus permitting spatial searches within the model embryo of, for example, the distributions of dividing cells or particular gene products. The mapped data can, of course, be separately organized in a conventional database, making it accessible to the power of a non-spatial database query system; an example is the link between the Atlas models

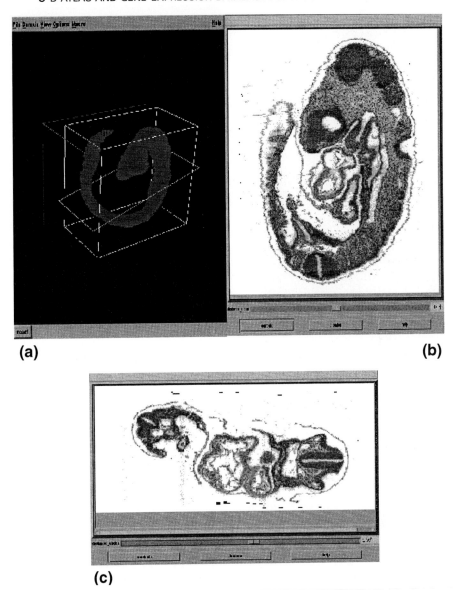

Figure 3. The Theiler Stage 14 (E9) model embryo of the Mouse Atlas viewed in the prototype image-segmentation interface. (b) A digital, parasagittal section through the Theiler Stage 14 model embryo (originally reconstructed from images of transverse wax histological sections). The voxels represent $4 \times 4 \times 7$ μm and the image displays detail at the tissue level. The neural tube and other major structures have been delineated (not shown). In (a) the neural tube of the same embryo is shown for purposes of orientation within a box bounding the complete 3-D model (the rest of the embryo is not shown). The user can digitally resection the reference embryo at any angle, using interactive controls, until the reference section shown in (c) matches, as closely as possible, the experimental section as observed under a microscope. The current plane of section is shown automatically in the orientation box in (a). The interface can then be used, with image-segmentation tools, to enter gene-expression data from the experiment onto the reference image, and thus spatially to map the data into the framework of the Atlas.

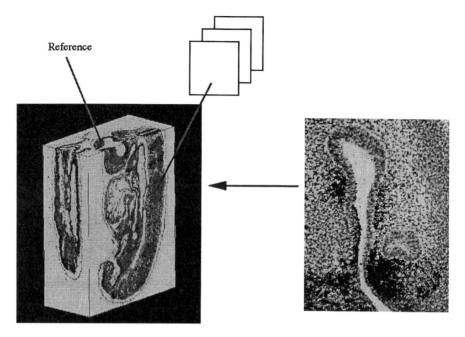

Figure 4. Spatially mapped data in the Mouse Atlas. Information can be mapped into the framework of the Atlas (left part of the diagram) in numerous ways. Text data (e.g. references), or image data (e.g. images of several examples of the phenotype of a particular mutant) can be mapped as a block to a domain in the model delineated by the user. In this way, for example, the 'domain of influence' of a mutation may be mapped and the database can be searched in spatial terms to find which mutations affect a specified part of the embryo, perhaps corresponding to the expression of a particular gene. Alternatively, images (represented in the right part of the diagram) can be mapped voxel by voxel onto the reference model to achieve full spatial mapping. The mapped data may be 3-D reconstructions or 2-D images, for example images of sections labelled by *in situ* hybridization to show gene-expression domains, as illustrated here.

and the Mouse Embryo Anatomy Database described below. The reference model can thus act as the focus for a very wide range of information about normal and aberrant embryo phenotypes.

Perhaps the most important information that must be mapped onto the histological images of the Atlas are the domains occupied by named anatomical structures, bones, muscles, blood vessels, and so on. This is being done by delineating the major named components of the embryo. The process is laborious (despite the use of image-analysis software) and involves careful study of the original sections under the microscope while 'painting' the domains. The anatomical domains, like any others, can be viewed simply in 2-D sections, or as 3-D surfaces visualized using commercial software such as AVS (Advanced Visual Systems Inc.) (*e-ref. 2*), which allows a wide range of manipulations, including stereo visualization using special spectacles. The benefits of seeing the 3-D relationships between structures and of viewing anatomical components in any combination, are obvious (*Figure 5*). The major reason for segmenting the image into named parts is, however, to create a

Figure 5. In order to relate the named anatomical components of the Mouse Embryo Anatomy Database to the 3-D space of the Atlas model, the major tissues have been delineated in the model and the neural tube, gut, and somites are shown here. Using this type of display, anatomical components and gene-expression domains can be viewed in any combination and rotated in 3-D view. The full, grey level reconstruction of the model embryo is hidden from view, but a low-resolution section is shown to illustrate how the 3-D view can be related to the more detailed, 2-D grey-level sections presented by the image-segmentation interface shown in *Figure 3*. The embryo reconstruction is visualized here using AVS software (*e-ref. 2*).

bridge between image and a text description of the embryo, thus combining the advantages of both formats.

Using the Mouse Atlas, researchers will be able to explore normal development of histological structure, for example, by comparison with mutant embryos and to delineate any region of interest and attach data to this region. Since the same reference models can be used by any number of researchers, it will be possible to share data and, in principle, to build databases that describe the embryo phenotype in some detail. In addition, the Atlas will help researchers without expert anatomical knowledge to carry out a number of essential practical tasks, for example to find the plane of section in an experimental embryo (*Figure 3*), and to identify the major structures, a first step in many histological analyses, including the interpretation of, for example, immunohistochemical and *in situ* hybridization experiments.

The Mouse Atlas is being built at the MRC Human Genetics Unit and the Anatomy Department, University of Edinburgh and will be available as a set of CD-ROMs. Initial versions will be for UNIX machines running Solaris 2.5, but basic functions will also be available for multi-platform access via the Word Wide Web (WWW) (*e-ref. 3*). The first model embryos (fertilization to 9 days postconception, Theiler Stage 14) will be available late in 1997 and older stages will be made available as they are completed.

2.2 Text: the Mouse Embryo Anatomy Database

To begin to establish a standard nomenclature for the anatomy of the mouse embryo, we are building a database of all morphologically recognizable, named structures at each of the 26 developmental stages defined by Theiler (Theiler, 1989). The names will be organized in a spatial hierarchy (*Figure 6*) with options for displaying alternative groupings of components. The database will include a thesaurus of alternative names, information on tissue type/architecture, and notes and references on contentious aspects of the nomenclature.

The database can, of course, also hold data on properties attributable to these named structures. It is planned, for example, to incorporate information on lineage relations between anatomical components. Other data can, in the future, be added to these anatomical names, for example bibliographic information, descriptions of development, and so on, indeed any information, in text or image form, that is best linked directly to a name rather than to a spatial domain in the Atlas. The organization of the database as a spatial hierarchy allows results, such as gene-expression patterns, to be documented at any resolution and additional detail to be added where required.

It is possible, and desirable, that this nomenclature will become established as a standard for diverse results relating to mouse development, thus anchoring a potentially very large amount of information to the Mouse Atlas; for example, as described below, the Anatomy Database is being used in a collaborative project to establish a database of gene expression for the mouse.

The Mouse Embryo Anatomy Database will be available on the WWW (*e-ref. 3*). Nomenclature up to Theiler Stage 22 (approximately E14) will be available in 1997, and later stages during 1998. The database is being built at the Anatomy Department, University of Edinburgh, and the MRC Human Genetics Unit.

2.3 The relative merits of text and image formats for mapping data to the Mouse Atlas

Text, as a widely used form of knowledge representation, integration, and interrogation, potentially provides links across an enormous range of data. Text can be used to name not only structures, but events, and to qualify and give attributes to

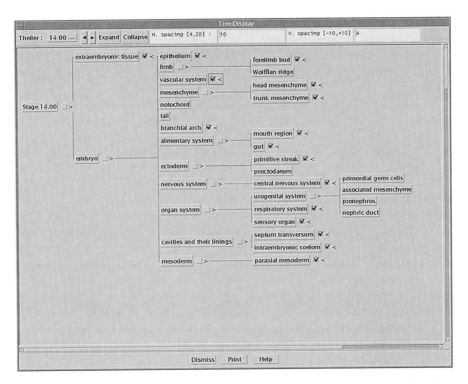

Figure 6. Named anatomical components in the Mouse Embryo Anatomy Database. Part of the prototype anatomy database is illustrated, showing some of the named components of the Theiler Stage 14 embryo (E9). The anatomical names are shown in 'tree' view to illustrate the hierarchical nature of the database. Clicking on any name calls up a window which allows the user to view anatomical data pertaining to that structure. Part of the tree is 'collapsed' (e.g. extra-embryonic tissue), part is expanded (e.g. embryo … pronephros). Different parts of the tree can thus be viewed at different resolution and the names used to describe, for example, gene expression domains.

data in ways that cannot be achieved by any other means that so readily permits human interaction. Providing that a standard vocabulary is used, moreover, digitally represented text can be very easily and quickly searched. For use in a database, not only must we have a standard vocabulary, but the *interpretation* of the terms must be defined unambiguously in order to ensure data integrity. These definitions will always be a source of discussion among anatomists, especially as knowledge grows (for example, from molecular experiments), but for database purposes, interpretation of existing terms must be fixed and new information added as *data*. To this end, because some of the boundaries between named structures are arbitrary, an explicit description of each term in space is required so that each can be interpreted without ambiguity, albeit not without argument.

The limitations of text become apparent, however, when one attempts to document certain developmental processes in the early embryo. One particular difficulty is encountered in describing gene-expression domains, since these do not necessarily correspond, in time and space, to those divisions of the embryo that have been recognized and named by classical anatomy (of the many possible examples, see, for instance, Cohn and Tickle, 1996; Davidson, 1995). When structures and cell types have begun to differentiate, gene expression can, of course, usually be assigned to named components. During the process of pattern formation and in the early stages of organogenesis, however, there is often a mismatch between gene expression and emerging structure. This is not merely an inconvenience, but reflects a fundamental feature of development. The genetic control of pattern formation is mediated by the spatial deployment of interacting signalling systems (see, for example, Cohn and Tickle, 1996; Lawrence and Struhl, 1996) and the appropriate format for describing gene-expression patterns is therefore spatial, that is, an image format. In contrast, the anatomical components that comprise the gross aspects of the embryo phenotype are recognized and named because they anticipate functional structures, generally of the adult body, limb, eye, vertebra, and so on. It is at the discontinuity between these spatial and textual descriptions of the embryo that many of the most interesting and important developmental processes are occurring and it is in attempting to bridge this gap that many investigations of gene function, in mouse and human, are focused. In order to investigate these processes, it will be necessary to bring together expression patterns and named anatomy within the same space. This is, of course, what is done in individual experiments by the techniques of *in situ* hybridization, immunohistochemistry and histochemistry (including reporter gene experiments). As discussed above, the Mouse Atlas can provide the appropriate framework for combining the results of many such experiments, using image or text formats as appropriate.

Even spatial domains of gene expression can, of course, be named, and it may become possible to use certain clearly defined and invariant molecular patterns as a 'molecular anatomy' to recognize new regions of the embryo and to add their names to the Anatomy Database. Similarly, critical molecular events during tissue development or cell differentiation (see, for example, Molkentin and Olson, 1996) can be used as temporal reference points for the expression of other genes and these temporal relationships are best documented in text (as a gene being expressed 'before, concurrent with, or after' the expression of the reference marker).

In addition to their suitability for different kinds of data, textual and image formats have practical differences. Using the graphical approach, it is not necessary to have identified parts of the embryo by name in order to enter data or query the database, a feature of considerable value to those with limited anatomical training. Moreover, because anatomical structures and expression domains may have complex shapes, certain significant spatial relationships within the data may only be perceived in 3-D graphical form. A disadvantage of the graphical approach is that each model embryo represents only one time point in the process of development. In contrast, the text-based approach can list all the anatomical components to be found during each stage of development and is thus more flexible in this respect. The complementary and integrated use of both formats may eliminate these disadvantages.

There is, however, an important caveat to the use of either text or graphical methods to record data. The resolution of the process of transforming different data into a common format may not be sufficient subsequently to allow exact comparisons of pattern. In many respects this limitation applies to the everyday comparison of experimental results reported in the literature; mapping and collating information from different sources in a database simply formalizes this process of abstraction. The database approach introduces additional difficulties, however, where important features of the data are blurred by the mechanisms of mapping and querying, or oversimplified by the coding format. While efforts can be made to avoid these difficulties, a gene-expression database is best regarded not as a representation of precise relationships, but as a 'hypothesis generator', enabling possible relationships to be perceived which must then be tested directly by experiment.

3. The Mouse Gene-Expression Information Resource

The Mouse Atlas is being developed as the format for a publicly accessible, international database of gene expression in the mouse embryo. This database will incorporate the MRC Graphical Gene-Expression Database, and a closely linked, text-based database of gene expression being developed at The Jackson Laboratory, the GXD. This configuration of modular tools will function as an integrated unit, the 'Mouse Gene-Expression Information Resource' (MGEIR) (e-ref. 4), in which the user will be able to enter, manipulate and visualize gene expression data interchangeably in text or image form.

3.1 The MRC Graphical Gene-Expression Database

This database will use the 3-D images of the Mouse Atlas as a framework to store spatially mapped *in situ* gene-expression data, that is, data from *in situ* hybridization, immunohistochemistry and histochemistry experiments (including reporter gene experiments). The database is still in the development phase: a prototype has been built, and we anticipate that a publicly accessible database will be available by 1999. Operational instructions will be published via the WWW and the general considerations that govern data preparation for any such database have recently been reviewed (Davidson *et al.*, 1997a). In this section, we describe, in outline, those features of design and operation relevant to the extension of this system to an electronic database of gene expression in the human embryo.

Access to the remote database will be via the WWW. The researcher will be able to sit at a computer in their own laboratory and enter data or interrogate the database using the Mouse Atlas as a local tool to help analyse their experimental results; for example, the user will be able to analyse their data in 3-D space, in relation to the expression of other genes and to the developing phenotype. This might involve interrogating the remote database with such queries as 'list all genes which are expressed close to (defined by the user) the domain of expression of the test gene in a defined part of the limb bud epithelium at day E10.5 of development', or, 'which anatomical components express genes A and B at E10 and gene C at E11?'. The user will also be able to record and store results and share them with similarly equipped laboratories during the course of a project (for example, with restricted access to collaborators). The same system will be used to submit data for publication in the public database.

To enter data for analysis or submission will require mapping the gene-expression pattern onto the model embryos of the Mouse Atlas. Two methods will initially be provided. Both operate on a section by section basis or on successive views of whole-mount preparations. The first will use custom-built software for manual image segmentation to define gene-expression domains, for example by examining each section of the original data under the microscope and drawing the pattern on the equivalent section in the appropriate reference embryo. The second method will use semi-automatic image mapping. Here, the operator will use custom-built database software linked to a conventional, digital camera mounted on a microscope to record images of the histological structure and the signal marking gene expression in each section. After finding the equivalent sections in the appropriate stage model embryo from the Mouse Atlas, the operator will mark a few equivalent positions in the histological structure of the experimental and reference images. Using these 'tie points', the software will transform the experimental image to match the reference image and the same transformation will then be used to map the signal onto the reference image. In practice, a combination of manual segmentation, semi-automatic image transformation and text description will be used; for example, if a gene is expressed in part of the neural tube, one may begin by entering 'neural tube': this will display the anatomical domain of the neural tube from the Mouse Atlas which can then be modified using the manual segmentation software to delineate quickly the actual domain of gene expression. For submission, the mapped data will be transmitted to the remote database. The MRC graphical database will be fully integrated with the GXD and submission of graphical data to the MGEIR will include, via the GXD, at least one original image (which is not spatially mapped) as well as ancillary information and any further textual descriptions of the gene-expression pattern. The remote database will also be accessible for queries and the results will be returned for local analysis in the environment of the Mouse Atlas.

The MRC Graphical Gene-Expression Database (*e-ref.* 5) is being developed at the MRC Human Genetics Unit and the Anatomy Department, University of Edinburgh.

3.2 The GXD Gene-Expression Database

Different expression assays provide different, but complementary, expression information. *In situ* studies provide detailed insights into spatial expression domains. Northern and Western blot analysis reveal the number and sizes of transcripts and proteins. Methods such as RNase protection and reverse transcription polymerase chain reaction (RT–PCR) can detect small amounts and small differences in transcripts. Expression assays using high density of cDNA clones or gene-specific oligonucleotide arrays can generate semi-quantitative expression information for thousands of genes.

GXD (*e-ref.* 6) will store and integrate data from the various types of expression assays. As new data and new assay types can be readily added, the database can provide up-to-date information about what products are known for a given gene and about where and when these products are expressed. Expression patterns are described in standardized text using the Mouse Embryo Anatomy Database. For *in situ* studies, the textual annotations are complemented by digitized images of original expression data and these images are indexed via the terms from the standard nomenclature. In this way, the expression information becomes accessible to global database queries.

GXD will be fully integrated with the Mouse Genome Database (MGD) (*e-ref.* 7). This will foster close links to genetic and phenotypic data of mouse strains and mutants, and comprehensive interconnections with other mammalian databases, such as the Genome Database, GDB (*e-ref.* 8) and the On-line Mendelian Inheritance in Man, OMIM (*e-ref.* 9). Numerous pointers to other relevant databases, including the major sequence databases, will place the gene expression information into the larger biological and analytical context.

3.3 The integrated GXD and MRC graphical gene-expression database

Access to the MGEIR will use an integrated interface to both graphical and text databases which enables the user transparently to combine text and image methods. Retrospective data from the literature will be entered by database curators. New data will be submitted mainly by individual researchers and checked by editorial staff. The MGEIR may also offer the option of peer review so that data can achieve full publication status.

The MGEIR is being developed with advice from numerous laboratories in Europe and North America with the aim of developing a user-friendly database system for the community at large. Currently, prototypes of the GXD submission interface are being tested by these sites. Although the prototype versions of the GXD interface currently run on Macintosh computers, it is intended that the integrated GXD and MRC gene expression databases will be accessible through platform-independent, WWW interfaces.

4. Challenges for a database of development

In spite of the force of argument in favour of a database of development, the approach faces considerable obstacles. The first, and most serious, is the effort required to enter data. In any database which relies on researchers to enter their own results, it is crucial that data entry provides tangible benefits and, above all, is simple. Some of these benefits will be immediate; for example, in organizing laboratory results or aiding the interpretation of experiments. Once the database contains sufficient information, the newly entered expression pattern can be analysed in relation to the expression of other genes. A further benefit is that the database will allow the researcher to publish purely descriptive results which may be unacceptable to conventional journals.

One way to reduce the work of data entry is, of course, to simplify the data, either by documenting results in less detail or by eliminating all but essential ancillary information. The Mouse Atlas and MGEIR will, of course, be able to hold much more comprehensive information than the present, conventional methods of publication. The MGEIR is, however, built on the principle that data can be entered at almost any resolution and, if required, with only sufficient information about gene, embryo and probe, to allow the experiment to be interpreted. Quality rather than quantity is the important criterion for accepting data and the only additional requirement for a minimum database entry will be a single image showing original data so that quality may be assessed.

The work involved in data entry should be viewed in proportion to the benefits to be gained in exploring gene function. It should also be borne in mind that graphical computer methods and hardware are evolving rapidly. Nevertheless, it is not yet clear if users will be persuaded to enter their own results or if the balance will shift away from local data entry towards the provision of a centralized data entry service by database staff funded by the community at large.

A further obstacle to the database approach is the incomplete nature of much of the information relating to embryonic development. This arises partly as a result of the *ad hoc* experimental approach, and partly because researchers rarely have time to document gene expression thoroughly. Although the same difficulties affect the interpretation of results published in the literature, attempts to codify data explicitly will bring into sharp focus any gaps in the observations. The solution adopted by the MGEIR is to store data in its original form as far as possible, and to deal with the problem of incomplete coverage later, when the database is searched: the search procedure will uncover all possible instances of expression in a specified region, but the results will be presented in such a way that the user sees the extent of the original observations.

Over the next decade, gene expression is likely to come from two very different kinds of study. Extensive screens will provide semi-quantitative information on an organ-by-organ basis, documenting the expression of many thousands of genes [indeed potentially almost all genes (Schena, 1996; Schena *et al.*, 1996)], many of which are unknown or partially characterized. Intensive studies will provide a much more complete description of the expression of small numbers of known genes at tissue and cell resolution. In the mouse, where material is not limiting, large-scale screens may also produce 'snapshots' of high-resolution expression, for example, from single embryo sections. The capability to integrate information from these two kinds of study will be a challenge for any gene-expression database. The modular combination of the GXD, the MRC Graphical Gene-Expression Database and the Anatomy Database will integrate data from homogenized material and *in situ* studies, in text and image, at low and high resolution, allowing these results to be viewed together while keeping the origin and resolution of the data clearly flagged.

Finally, there is the challenge of combining information from different databases in order to investigate gene function. For the mouse, the MGEIR will be linked to the MGD and major sequence databases, but additional links may be necessary, for example to databases of gene expression in particular organs such as kidney (*e-ref. 10*) and tooth (*e-ref. 11*), larger databases such as the transgenic animal database, TBASE (*e-ref. 12*; Jacobson and Anagnostopoulos, 1996), which holds information on transgenic animals, including phenotypic descriptions of mice carrying targeted mutations, and any future database of gene–product interactions. Since information relating to gene function will also come from studies on other organisms, it will also be important to link data across a much wider field, including, of course, mouse and human studies. The common goal is a 'distributed database' which can be used as a single resource, but is built by integrating several otherwise independent databases. Software that allows independent databases to communicate with a minimum of internal modification [for example, the Common Object Request Broker Architecture, CORBA (Object Management Group, see *e-ref. 13*)] may ameliorate the difficulties of interoperability at the technical level, but there remain key issues of data compatibility, which will only be addressed by coordinated efforts across the community.

5. Extending the paradigm of the Mouse Atlas and gene-expression database to a future database for early human development

Although, in the past, the human embryo has received much less attention than that of the mouse, this situation is changing, as a result of the availability of suitable material (Burn and Strachan, 1995) and the application of increasingly sophisticated techniques to gain maximum information from this limited resource. 3-D Reconstructions of embryos have recently been made by several groups (Machin, 1996; *e-refs. 14, 15*). Systematic gene-expression studies are becoming more feasible and, in the future, are likely to make a major contribution to the understanding of human gene function and the molecular basis of inherited disorders (Strachan *et al.*, 1997). To store and collate the results of this effort will require some form of electronic database. Practical considerations will encourage the use of databases, at least at the local level: in particular, the scarcity of the material will make it necessary to analyse the expression of several genes in each specimen and thus to integrate and collate laboratory data on different genes, sections and embryos. Taking the wider view, however, it is crucial that the biomedical community as a whole can access results, preferably via a single database.

What will be required of a database of early human development? In general terms, the requirements discussed in relation to the mouse apply equally well to the human embryo. Moreover, the mouse is the principal model for human development (for reviews, see Bedell *et al.*, 1997a,b and Darling, 1996) and there is, therefore, a strong requirement to integrate mouse and human embryo studies, not only at the level of individual projects, but also at the bioinformatics level. Certainly, data from the two species must be compatible. With this in view, human embryo projects at the Department of Human Genetics, University of Newcastle upon Tyne, and the Institute of Child Health, London, have established links with the Edinburgh mouse atlas and gene-expression database group. In addition, the limitations inherent in studying human embryos emphasize the benefits to be gained from a complementary approach to using the human and mouse databases.

Human embryo material for gene-expression studies is a scarce and valuable resource. Extensive gene-expression screens are likely to become possible using the material that is available and, in some cases, these will be carried out in parallel in human and mouse. However, high-resolution studies, which are expensive in material, will necessarily be much less comprehensive in the human than the corresponding studies in the mouse, in terms of both developmental stages and number of genes examined. Thus, high-resolution analyses which require a reasonably full representation of potentially interacting genes will need to enlist data on the expression of homologous genes in mouse embryos.

Gene-expression patterns in the human embryo often can give only circumstantial evidence of gene function because there is no possibility of a direct experimental test. Experiments in human cells *in vitro* and biochemical studies may offer alternative approaches, but for the foreseeable future complementary evidence from genetic and embryological experiments in the mouse will play the dominant part in shaping our understanding of human development.

Within these limitations, a human database is likely to be used in the same way as the mouse system, to scan for potential interactions between gene products and for the association of expression with developmental processes, as well as to search for candidate disease genes by attempting to match expression patterns with the disease phenotype. A database of human development must, however, also be able to compare descriptive studies of human and mouse embryos and relate these to experimental results, including mutant phenotypes, from the mouse. One way to achieve such a close integration between the two databases would be to extend the paradigm of the Mouse Atlas and MGEIR to the human embryo. From this viewpoint, it is clear that the integrated use of the human and mouse databases will have applications in several important areas in human gene function research.

5.1 The integrated use of human and mouse databases

The analysis of genetic elements responsible for the regulation of gene expression will be a major aspect of gene function studies. The very large number of gene-expression patterns which are being documented can be expected to play an important part in this analysis. These data will include high-resolution *in situ* patterns, semi-quantitative observations on the distribution of gene products *within* gene-expression domains (locations of maximum and minimum abundance and the direction of gradients), and data from large-scale screens. The use of gene expression databases to identify similar patterns of expression and the combined use of the mouse and human databases to examine the conservation of pattern, will provide important preliminary information as an adjunct to sequence analysis and experiments to investigate the activity of regulatory sequences — including human sequences — in transgenic mice. One interesting application in this area may be in the identification of regulatory sequences that direct stage-specific and cell-, or region-specific, expression in one or both species. Such regulatory sequences will become a valuable resource for the genetic analysis of mouse development, for example to target gene knock-outs to particular tissues (Tsien *et al.*, 1996), as well as for gene therapy. With a potentially large number of such elements directing expression in, for example, the brain, the gene-expression databases may have an additional role in documenting these reagents in relation to the expression of native genes in order to facilitate the design and interpretation of knock-out experiments.

Another application of the mouse and human databases will be to determine which aspects of human development differ markedly from the mouse and, sub-sequently, to support studies which aim to understand these aspects of human development and to account for species differences. One area where differences are already clear and which will be a major focus for research is the nervous sys-tem, in particular the brain. Consideration of the database requirements for studying this system illustrates some general problems. Descriptive studies, aimed at correlating gene expression and cell activities with the primary organi-zation of the brain, will use the functionalities that have already been described. In addition, however, it will be necessary to study in detail the migration and development of neuronal cells individually, or in small groups. In order to describe the development of individual cells in mixed populations, the database must be able to document the distribution of cells carrying markers for cell type as well as their gene-expression status or cellular activity (division, apoptosis, etc.). In order to combine observations made on different embryos, it will be nec-essary to adopt statistical methods to integrate these distributions and to detect any underlying patterns. Similar methods will be required to collate and compare the distributions of cells in equivalent regions of mouse and human embryos and in individuals carrying mutations. The development and application of these methods will provide an exciting challenge for the future.

Following the identification of genes responsible for a large number of devel-opmental disorders, mouse models of human developmental diseases will clearly be a major area of study in the future. One promising application of the Mouse Atlas database approach is the possibility of building selected *disease model data-bases* to act as a focus for the study of disease models in different laboratories, in the same way as the Mouse Atlas will provide a focus for studies of normal devel-opment. Such a disease model database would comprise a series of digital recon-structions of the affected parts of mutant mouse embryos, representing the development of the disease. These models could be used as the framework for phenotypic, cell biological, and gene-expression data and as a tool with which to analyse and model the progress of the disease and the effects of therapies, in rela-tion to the parallel description of normal development in the Mouse Atlas. In addition, close integration of the mouse and human databases would allow the disease model database to relate to data on the corresponding aspects of human development, thus permitting the validity of the model to be assessed against the background of possible differences between mouse and human development.

This brief discussion of the possible applications of the Mouse Atlas paradigm, while clearly both selective and speculative, serves to illustrate the potential of this approach in the future.

5.2 The integration of human and mouse databases

How might the integration of human and mouse databases of development be achieved? The first requirement is to ensure that the databases use definitions of developmental stages in human and mouse embryos that are sufficiently refined and closely matched to enable comparisons between the data. In this respect, the scheme for relating the broad stages currently in use (based on O'Rahilly and Muller, 1987 and Theiler, 1989) will require additional, and more closely specified subdivisions.

Secondly, in order to ensure that the textual descriptions in both databases are compatible, it will be necessary to use matched, controlled vocabularies. To meet this requirement, an anatomical nomenclature for the human, corresponding to the Mouse Embryo Anatomy Database, is being built at the Anatomy Department, University of Edinburgh.

Thirdly, it will be necessary to compare spatially mapped data from the two species. This can be achieved in a number of ways as described earlier in this chapter. The most comprehensive method will be to determine a series of reciprocal image transformations, mapping voxels in a model embryo of one species to voxels in the stage-matched model embryo of the other in order to effect the translation and comparison of appropriate data. Such a mapping would not, of course, be uniform; topological differences exist between the embryos (for example, in the cardiovascular system) and many-to-one mappings will compress, and thereby distort, some of the mapped data. Within these limitations, however, it may be possible to obtain a useful mapping that would allow one to compare data from the two species (flagged as human or mouse) in a single image. It may even be possible to devise an archetypal vertebrate model which could be used for speculative comparisons of spatially mapped data. Such a model could bring gene expression, cell lineage and mutant phenotype information from other vertebrates, including the zebrafish (Driever *et al.*, 1996; Haffter *et al.*, 1996; *e-ref. 16*), into the domain of human embryo studies.

Acknowledgements

It is a pleasure to thank our colleagues, Bill Hill, Christophe Dubreuil, Margaret Stark, Elizabeth Guest, Allyson Ross and Jane Quinn at the MRC Human Genetics Unit, and Mathew Kaufman, Jonathan Bard and Renske Brune at the Anatomy Department, University of Edinburgh. The Mouse Atlas and MRC Graphical Gene Expression Database discussed here is the collaborative work of the MRC and University of Edinburgh groups and has benefited from the advice of our colleagues in the European Science Foundation Network on Gene-Expression Databases. The Mouse Gene-Expression Information Resource is a joint collaboration with the GXD team at The Jackson Laboratory, Bar Harbor, USA, and it is a pleasure to thank Martin Ringwald and his colleagues at The Jackson Laboratory for their help with sections of the text dealing with the GXD.

References

Electronic references (e-refs)

1. http://wwwcivm.mc.duke.edu/civmPeople/SmithBR/brs.html
2. http://www.avs.com
3. http:// genex.hgu.mrc.ac.uk/MouseAtlas
4. http://genex.hgu.mrc.ac.uk/ and http://www.informatics.jax.org/
5. http://genex.hgu.mrc.ac.uk/
6. http://www.informatics.jax.org/gxd.html
7. http://www.informatics.jax.org/
8. http://gdbwww.gdb.org/

9. http://www3.ncbi.nlm.nih.gov/omim/
10. http://www.ana.ed.ac.uk/anatomy/kidbase/kidhome.html
11. http://honeybee.helsinki.fi/toothexp/
12. http://www.gdb.org/Dan/tbase/tbase.html
13. http://www.omg.org/
14. http://magenta.afip.mil/embryo
15. http://www.st-and.ac.uk/~www_sbms/terrapin/frontpage.html
16. http://zfish.uoregon.edu/

Conventional references

Baldock R, Bard J, Kaufman M, Davidson D. (1992) A real mouse for your computer. *BioEssays* **14:** 501–502.

Bedell MA, Jenkins NA, Copeland NG. (1997a) Mouse models of human disease. Part I: Techniques and resources for genetic analysis in mice. *Genes Dev.* **11:** 1–10.

Bedell MA, Largaespada DA, Jenkins NA, Copeland NG. (1997b) Mouse models of human disease. Part II: Recent progress and future directions. *Genes Dev.* **11:** 11–43.

Brown SDM, Peters J. (1996) Combining mutagenesis and genomics in the mouse — closing the phenotype gap. *Trends Genet.* **12:** 433–435.

Burn J, Strachan T. (1995) Human embryo use in developmental research. *Nature Genetics* **11:** 3–6.

Cohn MJ, Tickle C. (1996) Limbs: a model for pattern formation within the vertebrate body plan. *Trends Genet.* **12:** 253–257.

Darling S. (1996) Mice as models of human developmental disorders: natural and artificial mutants. *Curr. Opin. Genet. Dev.* **6:** 289–294.

Davidson D. (1995) The function and evolution of *Msx* genes: pointers and paradoxes. *Trends Genet.* **11:** 405–411.

Davidson D, Baldock R, Bard J, Kaufman M, Richardson JE, Eppig JT, Ringwald M. (1997a) Gene expression databases. In: *In situ hybridization. A Practical Approach.* (ed. D Wilkinson). IRL Press, Oxford (in press).

Davidson D, Baldock R, Bard J, Brune R, Dubreuil C, Hill W, Stark M, Quinn J, Kaufman M. (1997b) The mouse atlas and graphical gene expression database. *Semin. Dev.* (in press).

Driever W, SolnicaKrezel L, Schier AF *et al.* (1996) A genetic screen for mutations affecting embryogenesis in zebrafish. *Development* **123:** 37–46.

Haffter P, Granato M, Brand M *et al.* (1996) The identification of genes with unique and essential functions in the development of the zebrafish, *Danio rerio. Development* **123:** 1–36.

Jacobs RE, Frazer SE. (1994) Magnetic resonance microscopy of embryonic cell lineages and movements. *Science* **263:** 681–684.

Jacobson D, Anagnostopoulos A. (1996) Internet resources for transgenic or targeted mutation research. *Trends Genet.* **12:** 117–118

Kaufman M. (1992) *The Atlas of Mouse Development.* Academic Press, London.

Lawrence, PA, Struhl G. (1996) Morphogens, compartments, and pattern: lessons from *Drosophila? Cell* **85:** 951–961.

Machin GA. (1996) Computerized graphic imaging for three-dimensional representation: general principles and application for embryo/fetal development. *Int. Rev. Exp. Pathol.* **36:** 1–30.

Molkentin JD, Olson EN. (1996) Defining the regulatory networks for muscle development. *Curr. Opin. Genet. Dev.* **6:** 445–453.

O'Rahilly R, Muller F. (1987) *Developmental Stages in Human Embryos.* Carnegie Institution of Washington, publication 637.

Ringwald M, Baldock R, Bard J, Kaufman M, Eppig JT, Richardson JE, Nadeau JH, Davidson D. (1994) A database of mouse development. *Science* **265:** 2033–2034.

Schena M. (1996) Genome analysis with gene expression microarrays. *BioEssays* **18:** 427–431.

Schena M, Shalon D, Heller R, Chai A, Brown PO, Davis RW. (1996) Parallel human genome analysis: microarray-based gene expression monitoring of 1000 genes. *Proc. Natl Acad. Sci. USA* **93:** 10614–10619.

Strachan T, Abitbol M, Davidson D, Beckman JS. (1997) A new dimension for the genome project: towards comprehensive expression maps. *Nature Genetics* **16:** 126–130.

Theiler K. (1989) *The House Mouse: Atlas of Embryonic Development* (2nd printing). Springer-Verlag, New York.

Tsien JZ, Chen DF, Gerber D, Tom C, Mercer EH, Anderson DH, Mayford M, Kandel ER, Tonegawa S. (1996) Subregion- and cell type-restricted gene knockout in mouse brain. *Cell* **87**: 1317–1326.

Index

Academic Services
Library Services